Horace Benge Dobell

On Coughs, Consumption, and Diet in Disease

Horace Benge Dobell

On Coughs, Consumption, and Diet in Disease

ISBN/EAN: 9783337035129

Printed in Europe, USA, Canada, Australia, Japan

Cover: Foto ©berggeist007 / pixelio.de

More available books at **www.hansebooks.com**

ON

COUGHS, CONSUMPTION,

AND

DIET IN DISEASE.

BY

HORACE DOBELL, M.D., F.R.M.C.S., Etc.

CONSULTING PHYSICIAN TO THE ROYAL HOSPITAL FOR
DISEASES OF THE CHEST, LONDON; PHYSICIAN
TO THE ROYAL ALBERT ORPHAN
ASYLUM, ETC., ETC.

PHILADELPHIA:
D. G BRINTON, 115 S. SEVENTH ST.
1877.

NOTICE BY THE EDITOR.

THIS work is made up of a series of extracts so arranged that they form a connected treatise on the Diagnosis and Treatment of some of the most common Diseases of the respiratory organs. These extracts have been drawn from the various published lectures of Dr. HORACE DOBELL, of London, one of the most accomplished physicians of our day, and one whose clear style, large experience and thoroughly practical mind peculiarily fit him to be an instructor in the delicate yet indispensable refinements of Physical Diagnosis.

The extreme frequency and the grave character of pulmonary and bronchial diseases render their study of the first importance to every practitioner; and the editor believes that it would be difficult to point out a writer from whom more of solid use concerning them can be learned than from the one here presented. In the difficult task of selection from the material at hand practical utility has been chiefly kept in view. Further than an occasional paragraph to indicate the connections, no additions to the author's text have been attempted.

In presenting the work, a brief biographical sketch of the author will certainly not be unwelcome to the reader.

Dr. HORACE DOBELL was born in the city of London,

Jan. 1, 1828. At the early age of fourteen he commenced his medical studies with Mr. Edward Fricker, F. R. C. S., with whom he spent ten hours a day for three years at professional studies, continuing his general education with private tutors at night. At this time he made a series of life-size drawings of the skulls of various races, which was long preserved in a neighboring museum. In 1845 he entered St. Bartholomew's Hospital, and during his pupilage commenced his contributions to medical literature by a striking paper "On the transmission of Disease from Parent to Offspring."

Having carried off several prizes for his essays, he became Member of the Royal College of Surgeons in 1849, and in 1856, Member of the Royal College of Physicians and M.D. of the University of St. Andrews.

For sixteen years he was Physician to the Royal Hospital for Diseases of the Chest, and is at present Consulting Physician to the same institution. He is also Physician to the Royal Albert Orphan Asylum, and of a number of other institutions and societies.

His literary activity has been great, and, in spite of the time demanded by his extensive practice, hardly a year has passed that some carefully written essay, lecture or treatise has not emanated from his pen.

The following list embraces the principal of these, without aiming at completeness:

1853. "On a New Means of applying Heat, and of maintaining the Temperature of Warm Applications." (*Abernethian Society.*)

1853. "On the Examination of Lives for Assurance." (*Medical Times and Gazette.*)

1858. "Demonstrations of Diseases in the Chest, and their Physical Diagnosis, with 36 colored illustrations."

1858. "Explanations of Amphoric and Tympanitic Resonance." (*Medical Times and Gazette.*)

1863. "A Contribution to the Natural History of Hereditary Transmission." (*Transactions of the Royal Medico-Chirurgical Society.*)

1864. "On Diet and Regimen in Sickness and Health." (6th Edition, published 1875.)

1864. "On the Germs and Vestiges of Disease, and on the Prevention of the Fatality of Disease." (Lectures delivered at the *Royal Chest Hospital.*)

1864-65-66. Results of Original Experiments with Pancreatic Juice, in a series of papers "On the Assimilation of Fat in Consumption," and "On the Treatment of Consumption with Pancreatic Emulsion."—Invention and Discovery. (*Lancet.*)

1866. "Lectures on Winter Cough, Catarrh, Bronchitis, Emphysema, Asthma."—based on a paper read at the Royal Medico-Chirurgical Society. (3d Edition, published 1875.)

1866. "On Tuberculosis; its Nature, Cause and Treatment." Setting forth an original hypothesis as to the nature of Consumption. (2d Edition, published two months after the first.)

1867. "On the True First Stage of Consumption." (Lectures delivered at the *Royal Chest Hospital.*)

1868. "On the Special Action of the Pancreas on Fat and on Starch."—Original Experiments. (*Royal Society.*)

1869. First Volume of "Dr. Dobell's Reports on the Progress of Practical and Scientific Medicine in different parts of the world," edited by Dr. Dobell, assisted by numerous and distinguished Coadjutors.

1870. Second Volume of these Reports.
1872. "On Affections of the Heart and in its neighborhood: Cases, Aphorisms, and Commentaries."—Illustrated.
1874. "On the Importance and Dangers of 'Rest' in the Treatment of Pulmonary Consumption." (*British Medical Journal.*)
1874. "A Contribution to the Natural History of Pulmonary Consumption: being an Analysis of 100 Male Cases of Hæmoptysis." (*Transactions of Royal Medico-Chirurgical Society.*)
1875. "Dr. Dobell's Annual Reports on Diseases of the Chest." Volume I.

The "Annual Reports" have achieved a deservedly high reputation, and it is hoped will be long continued.

CONTENTS.

	PAGE.
Notice by the Editor	3
Table of Contents	7–8

PART I.—The Diagnosis of Bronchial and Pulmonary Diseases.

I. The Systematic Examination of the Chest . . .	9
II. The Diagnosis of Early Phthisis	37
III. The Value of Cavernous Sounds	42
IV. The Importance of Hemoptysis as a Symptom . . .	45
V. Winter Coughs.—The Relations of Bronchitis and Emphysema	49
VI. The Diagnosis of Narrowed Air-Passages . . .	78
VII. Post-Nasal Catarrh	89
VIII. Ear Cough	97
IX. The Natural Course of Neglected Cough	112

PART II.—The Treatment of Colds, Coughs and Consumption.

I. Pathological Conditions in Winter Cough	121
II. The Early Treatment of Catarrh	132
III. The Avoidance of Colds	142
IV. Therapeutic Resources in Coughs	150
1. Medicines introduced by the Stomach . . .	151
2. Medicines introduced by Inhalation . . .	162
3. Counter-irritants	169
4. Changes of Climate	172

		PAGE
V. The Treatment of Post-Nasal Catarrh	. . . .	182
VI. The Management of Consumption	. . . .	185
Rest in Consumption	. . . . .	186
Pancreatic Emulsion in Consumption	. . .	190

PART III.—THE PRINCIPLES OF DIET IN DISEASE.

General Rules for Diet in Sickness	. . . .	193
Diets for Consumptives	. . . . . .	205
Diet for Diabetics	. . . .	207
The Use of Nutritive Enemeta in Disease	.	209
Special Recipes for Medical Food	. . . . .	211

PART I.

THE DIAGNOSIS OF BRONCHIAL AND PULMONARY DISEASES.

I. THE SYSTEMATIC EXAMINATION OF THE CHEST.

Inspection and topography of the Chest—Systems of Dr. Davies, Dr. Sibson, Dr. C. J. B. Williams—Objections to these and others—Importance of a basis on universal facts, such as the obvious anatomical points of the body—Objection to numerous instruments at the bedside—Simple guides in the practice of Inspection—Mensuration—Palpation—Auscultation and Percussion—Means of most easily acquiring a knowledge of these—Sources of typical sounds—Directions for experiments on inanimate bodies to educate the ear—Suggestions for facilitating correct diagnosis.

EVEN the most precise arithmeticians, when called upon for sudden mental calculations, are obliged to adopt some plan of "*ready reckoning;*" and something like this assistance is also needed at the bedside—where we are so often called upon to make sudden and important deductions—that we may seize upon the diagnostic features of disease, and place landmarks for the separation of group from group. To do this well is one of the chief acquirements of experience, and can be thoroughly learned only by the observation of disease; but I hope to be able to facilitate its acquirement by the following suggestions, and by giving some directions by which to smooth the path of clinical investigation into the physical signs of disease within the chest.

And first of Inspection.

Observation of changes in the form or movements of the chest is of great use in diagnosis, because it may, at a glance, lead the mind to an appropriate set of ideas relative to the seat of disease, and thus facilitate and give point to the further examination of the patient. This is within the proper sphere of inspection; it should be one of our first guides in a careful physical investigation—as the traveler on first entering a new locality takes a general survey of the place to learn the cardinal points, and the general bearings of the various pathways, before proceeding on any one of them for more minute exploration.

It is certainly possible to prosecute inspection with such analytical detail as to form, in some cases, a fair diagnosis from it alone, as it is also possible for a musician to play a whole symphony upon one chord of his instrument; but this sort of "harping upon one string" is what I would particularly point out for avoidance, as a very hazardous practice, subject at any moment to break down when least suspected, and at the most critical junctures. *A safe diagnosis must be based upon the conjunction of evidence from many sources.* I shall not, therefore, enter into all the details which can be worked up on the subject of inspection, but rather attempt to lay down a few common-sense directions for guidance in practice.

It has been the custom for teachers of auscultation to adopt or to contrive some plan of mapping out the chest-surface into regions and spaces, and to give names to each of these. Of such arrangements it will be sufficient to give the following examples:

Dr. Herbert Davies prefers the topographical arrangement designed by his father.

1. The clavicular region { Sternal. Middle. Humeral.
2. The anterior-superior region.
3. The superior mammary.
4. The submammary.
5. The axillary.
6. The superior-lateral.
7. The inferior-lateral.
8. The supra-spinal.
9. The infra-spinal.
10. The inter-scapular.
11. The dorsal, to which he adds
12. The supra-clavicular.

Dr. Sibson has made a very concise mapping, thus :—

The simple regions,
- Right pulmonic.
- Left pulmonic.
- Cardiac.

The compound regions,
- Pulmo-hepatic.
- Pulmo-gastric.
- Right pulmo-cardiac.
- Left pulmo-cardiac.
- Pulmo-vasal.

And Dr. C. J. B. Williams gives a somewhat different plan from either of these.

1. Clavicular (subclavian of Laennec).
2. Infra-clavicular (anterior-superior, Laennec).
3. Mammary.
4. Infra-mammary.
5. Superior sternal.
6. Middle sternal.
7. Inferior sternal.
8. Axillary.
9. Lateral.
10. Inferior-lateral.
11. Acromial.
12. Scapular.
13. Inter-scapular.
14. Inferior dorsal.

Many arrangements may be found, each differing,

in some respects, from the rest; and any one of these may be adopted by those who desire it, but in my opinion they had better all be rejected. The simple fact that there are so many arrangements destroys the use of any one—because the only object in making them must be to facilitate the accurate description of the site of disease. But as there are numerous plans of topography, the divisions of which differ in their limits, no accuracy of description can be obtained by their use unless we first state, in each case, the author of the arrangement followed; and then it will be necessary to give the definitions, otherwise it will only be intelligible to those who are acquainted with the same author; and at last, some further method of division must be adopted, to indicate how much, and what portion, of any region is affected.

It is of the greatest importance that the terms of science shall be clearly intelligible to as large a number of persons as possible; and hence, whenever it is practicable, they should be based upon universal facts. In the present instance, we have a basis of this kind in the *obvious anatomical points*, such as the clavicle, its two ends and middle; the scapula with its angles and spine; the sternum with its divisions; the ribs with their numbers, their angles, their cartilaages, etc.; the axillæ; the nipples; the intercostal spaces with their numbers; and the spaces above and below the bones mentioned. These anatomical points, which are familiar to all, may be taken as the landmarks from and to which measurements may be made, with any degree of accuracy desired.

And by having a scale of inches marked upon the ear-piece or some other part of the stethoscope, a gauge is always at hand. Other apparatus should only be *kept in reserve for special occasions*, to test the accuracy of more simple guides; to see a physician at the bedside constantly accompanied by a hammer, a pleximeter, and a chest-measurer, besides his stethoscope, is not only most alarming to his patient, but a pitiable admission of clumsiness in the use of natural gifts.

To proceed then with some simple guides to inspection. The chest may be variously deformed from causes independent of disease of the lungs or pleura. Setting these alterations aside: *if one portion of the chest is depressed*, it is due either to disease of the lung beneath, causing diminished power of expansion, and obliging the chest-wall to sink down upon it, as in tuberculous excavations; or to compression of the lung from some cause external to itself—most probably in the pleura—the depressing cause having disappeared, and the chest-wall being allowed to apply itself to the diminished lung, as after empyema.

If one portion of the chest is enlarged, it is due either to increased function of the enclosed lung, compensatory to loss of function at some other part; or to diseased increase in the size of the enclosed lung, as in emphysema; or to an accumulation of some abnormal matter in the chest, as in the case of intra-thoracic tumor, empyema, hydrothorax, or pneumothorax. When this matter is in the pleural sac, its specific gravity will influence its position.

MOVEMENTS OF THE CHEST.

The normal respiratory movements, when carefully examined, during ordinary quiet breathing and during forced expiration or inspiration, illustrate the principle deviations met with in disease. They are well shown in the accompanying figures (see figs. 1 and 2, page 16), first designed by Mr. Hutchinson (vol. xxiv, 'Royal Med. Chir. Trans.'), in which the outer black line indicates ordinary respiration in the two sexes, the thickness of the line representing the extent of movement; the inner black line forced expiration, and the dotted line forced inspiration.

It is here seen that in health the *thoracic* movements predominate in the female, the *abdominal* in the male. If, therefore, the breathing movements in a man should have the character of those normal to a woman, they would in him indicate disease, and be called "thoracic respiration." If, on the other hand, the breathing movements of a woman should have the character of those normal to a man, they would in her indicate disease, and be called "abdominal respiration."

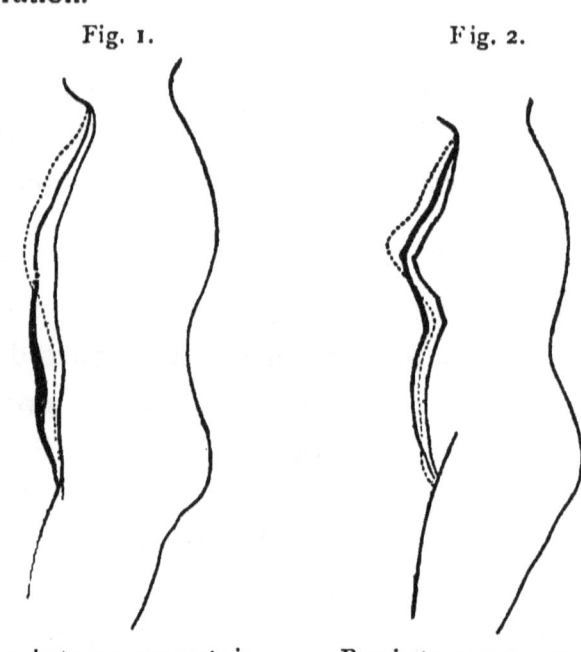

Fig. 1. Fig. 2.

Respiratory movements in the male.   Respiratory movements in the female.

Again, it is shown by the figure that in both sexes the movement of forced inspiration predominates in the thoracic portion, as indicated by the dotted line, and that, while the chest-wall is elevated, the abdomen actually recedes. This may be particularly observed when the respiration is rendered forced and laboring by disease.

MOVEMENTS OF THE CHEST IN DISEASE.

If the thoracic breathing acts are in excess, the disease is one either interfering with the action of the diaphragm, as in peritonitis; or in which the difficulty of getting air into the lungs is so great that all the muscles of the chest-wall are exerted, to attempt—by elevating it and taking off the pressure from the lungs—to induce the further expansion of those organs and the consequent ingress of air, as in some cases of bronchitis, spasmodic asthma, and the like.

If the abdominal breathing acts are in excess, the disease is one requiring the diaphragm to compensate, by excess of action, for deficient action in the muscles of the chest,—for loss of the healthy resiliency of the lung,—or for both.

If the breathing acts are one-sided, it may be due to deficiency on one side, or to excess on the other, or to both,—the excess compensating for deficiency; and these affections may be in the chest-wall, as in intercostal neuralgia, in the pleura, or in the lung.

If the expansile action is deficient but elevation remains, the loss of power is in the lung, but it may be due to disease of its own structure, as in phthisis, or to pressure from without, as in consolidation from the pressure of interpleural fluid.

If the expiratory movement is undue in length, the

disease is one offering an impediment to the escape of tidal air from the lungs, which may be due to actual obstruction in the tubes, or to loss of power on the part of the lung to contract upon the air and expel it. Thus it occurs in bronchitis, in tubercle, and to its greatest extent in emphysema.

If the respiratory movements are excessive in rapidity, it may be due to any disease preventing the full expansion of the lungs at each inspiration, whence an attempt is made to compensate the loss of quantity by increased rate; or to imperfectly aërated blood, or to any disease increasing the rapidity of the circulation, whence an attempt is made to supply air to the blood at a rate proportionate to the demand.

If both thoracic and abdominal breathing movements are deficient, it is due to a combination of diseases interfering with the action of the diaphragm and the external muscles of the chest, or to paralysis, or most probably to general exhaustion, in which case the deficient action will be broken at intervals by a sighing inspiration, as sufficient power accumulates for the effort.

MENSURATION.

When inspection of the chest indicates any abnormal character, the amount of deviation may be estimated by *mensuration*, or if the eye fails to detect any deviation from the state of health, the correctness of inspection may be tested by the rule; hence in minute examinations mensuration must form a part.

PALPATION.

Palpation of the chest is only one of the many instances by which the importance of educating all the special senses is illustrated in medical practice. By it we may learn, among other matters—

1. The amount of resistance in percussion offered to the percuting finger.

2. The degree to which the vibrations (*fremitus*) of the thoracic organs are influenced by disease, and to which the transmission of these is increased or diminished by interposing media.

3. The presence of fluid, indicated by fluctuation.

4. The comparative temperature of different parts of the surface, and the like.

None of these means of acquiring knowledge is to be neglected, and the relation of what is learnt from one to that obtained by another is especially deserving of attention.

AUSCULTATION.

Theoretically, it is of little consequence what terms are chosen to identify certain things, provided that their meaning is clearly understood by those by whom, and for whom, they are used. But in the practical application of any science or art, it is of the greatest consequence to use as few terms as possible, and to choose those which are plain, and easily remembered. In this work this fact has been kept in view; and, therefore, instead of selecting terms because they might be most euphonious, or because their etymology proved to be most consistent with their application to particular things, I have chosen those which I find most familiar in the mouths of men engaged in the practice of correct auscultation, and which, I believe, most students will find retaining a place in their own vocabularies after a few years of active public or private practice.

They are few in number and most of them short, thus:

1. Respiratory murmur
   - Inspiratory.
   - Expiratory.
     - Harsh.
     - Weak.
     - Jerking.
     - Supplemental.

2. Bronchial
3. Cavernous
   } breathing, cough or voice.

4. Rhonchus
5. Sibilus
   } Used for "sonorous rhoncus" and "sibilant rhonchus."

6. Crepitation
   - Large
   - Small
   } Used for "subcrepitant rhonchus."

7. Fine crepitation, used for "crepitant rhonchus."
8. Gurgling.
9. Friction sound, of various characters.
10. Gutta cadens, commonly called metallic tinkling.
11. Metallic tinkling   do.   do.
12. Amphoric echo.
13. Bruit de pot' fêlé. ⎫
14. Ægophony.   ⎬ Retained only because they have become familiar as names for certain sounds.
15. Succussion sound.
16. Percussion sound, normal.
17.   "   "   impaired.
18.   "   "   dull.
19.   "   "   amphoric.
20.   "   "   tympanitic.
21.   "   "   metallic.
22. Vocal fremitus.
23. Rhonchal fremitus.

Any further varieties in the general characters, the pitch, the timbre, the reciprocation, &c., of these sounds, may be described in appropriate terms by the observer, and should be mentioned as something supplemental to the more usual characteristics.

1. Learn to be perfectly certain that a respiratory murmur is healthy; that is, that it comes from vesicular lung-substance in its healthiest state, and is conducted to the chest-wall through structures in their healthiest state. You will then be able to say, what

few physicians can venture upon, "There is no disease in this portion of lung which physical diagnosis is at present capable of detecting." There is only one means by which to acquire this knowledge, viz., listen to the healthy chest until its sounds are indelibly impressed upon the mind. For vesicular respiratory murmur, listen below the centre of the clavicle in front, and beneath the inferior angle of the scapula behind. Do this in an infant, a child of five to ten years old, an adult man, an adult woman, an old man, an old woman. In each case examine a thin subject and a fat one; and in each case, in immediate association with the respiratory sounds, examine those of the voice, cough, forced breathing, whispering, and percussion. After each examination think carefully over what you have heard, and return again to examine and to think until you can call up in the mind, at will, the characters of the sounds you have heard. Attach especial value to their *essentially individual* characteristics.

Should this appear to any one too severe a lesson, let him give up the idea of ever becoming a competent auscultator, for sooner or later he will find that there is no shorter or easier way of arriving at that end; and it will be less tedious and less disappointing to begin at first upon the sure road. It is astonishing how much you have learned, negatively and positively, when you know healthy vesicular sounds under their typical varieties. It is the most delicate point in the whole practice of auscultation.

2. Impress clearly on the memory the breathing, cough, whisper, and voice sounds of the trachea; which may be heard through the stethoscope placed over any portion of that tube; and remember that these sounds are typical of a cavity of considerable size, with walls of a certain degree of smoothness and hardness.

3. Familiarize the ear with the sounds of breathing, cough, whisper, and voice, in a bronchial tube; and especially with the distinctive characters of these compared with the vesicular and tracheal sounds. This may be done, in most healthy chests, by following with the stethoscope the course of the trachea down to its bifurcation (see fig. 3), and listening, as you proceed, to the change of sound in passing from the main tube to either of its branches; and still advancing, watch the change as the bronchus becomes imbedded more deeply in vesicular lung-substance. When you have lost the bronchial, and hear only the vesicular sounds, place the stethoscope over the back of the chest, between the scapulæ, in a line with the second and third dorsal vertebræ, where again you will recognize the bronchial sounds. (In a few healthy chests bronchial sounds cannot be heard in these regions.)

4. Particularly bear in mind why you lost the bronchial sounds, which though lost did not the less exist, but were simply inaudible at the surface of the chest, when vesicular air containing lung-substance intervened.

Fig. 3.

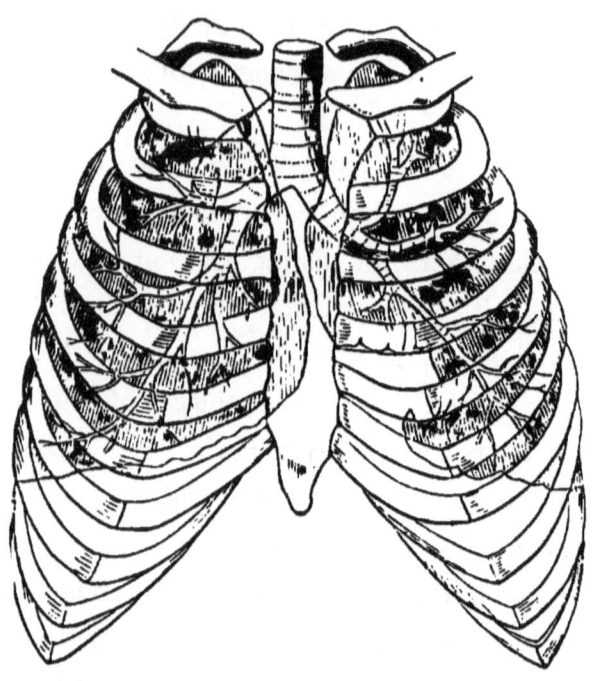

Position of the lungs midway between expiration and inspiration, and the relations of the trachea and principal bronchi to the ribs and sternum.

5. Remember that healthy vesicular murmur *cannot be imitated* by anything, but must always come from healthy lung; but do not forget that as obscures the sounds which you know are passing

beneath it in the healthy bronchi, so may the sounds of disease be going on beneath it and yet be inaudible; and remember also that in its turn vesicular murmur may exist, but not be heard.

6. Remember that tracheal sounds always indicate disease when heard elsewhere than over the trachea—that in health there is no cavity in the lung large enough to yield them after you have passed the first bone of the sternum.

7. Remember that bronchial sounds always indicate disease when *heard* elsewhere than at the two points, in which you have already found them in the healthy chest; but do not forget that they *exist* throughout the whole bronchial tree, and need only a change in the conducting medium to bring them audibly to the surface.

8. Familiarize the ear with the breathing sounds through your own nose during slow, quick, and forced respiration, and with all the varieties you can produce by different degrees of compression applied to the nares; also take notice of the nasal breathing sounds when you are suffering from catarrh. For many of the sounds to be heard in the chest, due to alteration in the calibre and in the mucous lining of bronchial tubes, may be thus learned, with the advantage of being able to identify the cause of each.

Educate the ear by experiments on inanimate bodies.

Take a piece of vulcanized rubber tubing, three feet long and two lines in the bore: with this a number of useful lessons may be learned. Rest about half the tube on the table or against a wall and apply a stethoscope upon it lightly; listen, while, with one end of the tube in the mouth, you inspire and expire through it alternately. The sounds heard will be something like loud bronchial breathing. Make pressure upon one portion of the tube so as to diminish its calibre at that part, and if it is perfectly dry, *sibilant* sounds will be produced when the tube is most compressed, and sonorous snoring *rhonchus* when the calibre is only slightly diminished. By varying the extent and degree of compression, changes in the character and pitch of the breathing sounds may be produced, as numerous as those heard in the chest. It is most interesting to observe the effect of moisture in the tube, and how minute a quantity affects its sounds. If sixty drops of water are put into the three-foot tube, and the auscultation and breathing repeated as before, *large crepitation*, most abundant and loud, will appear to fill the tube for a considerable part of its length. If pressure is now made, as for the production of rhonchus and sibilus, the size of the crepitation will diminish as the calibre of the tube is decreased. If we empty the tube and blow through it to expel all the moisture possible, and again auscult it, crepitation will be heard, not of the same character as before, but more crackling, yet still distinctly bubbling; and if the tube is compressed the sounds become smaller, and assume nearly the character which they had before the water was emptied out. At parts where the tube appears dryer, loud rhonchus is heard, less musical and harsher than from the dry compressed tube. So slight an amount of moisture is sufficient to give the crepitation and rhonchus, that it is most difficult, having once moistened the tube, to get it again so dry as not to produce these sounds: so that we cannot help wondering how it is that the human air-tubes are ever free from rhonchi and crepitations, when lubricated by even their normal amount of secretion.

Take a vulcanized rubber ball, with a hole in one side, partly fill it with water, and introduce, through the hole, a tube with its extremity beneath the surface of the fluid. By listening through a stethoscope placed on the ball while we breathe through the tube, *gurgling* will be heard in the ball, varying in its character with the force of breathing, the size of the tube, and the depth of the fluid through which the air has to pass. The smaller the aperture of the tube the more the gurgling loses its distinctive characters and approaches to *crepitation*.

If a *dry* ball and tube are now taken and ausculted again during respiration, breathing sounds of cavernous character are heard. These, however, are best learned by ausculation of the trachea; but a point of interest may be noticed, viz., that the sounds of breathing and voice in the ball appear to penetrate the stethoscope much more than the same sounds traversing a large tube. This may be demonstrated by ausculting, first, the ball; then a vulcanized rubber tube, half an inch in the diameter of its bore.

The sound of *succussion* is well imitated by filling a water cushion partly with air and partly with water, and then either jolting it in such a manner as to dash the water against the wall of the cushion, or shaking it so as only to splash the water upon itself. The sound will be more gurgling in the latter than in the former case.

If a glass bell is suspended in air and then a small small blast of wind suddenly directed against its margin, we may hear, first, the rush of air, and then, following it at a distinct interval, a faint ring of the bell; the same effect is produced if the gust is directed against the interior wall of the bell. This corresponds in essential characters with the sound called *metallic tinkling*.

If we now suspend lightly a thin glass or metal flask, so that it can ring when struck; and while listening attentively, with the ear close to, but not touching, the wall, pour some water slowly, drop by drop, into it from the neck, a dropping is heard accompanied by a slight ringing, but quite distinct from the sound of the last experiment. This dropping corresponds in essential characters with the sound called *gutta cadens*. As the water collects in the flask the dropping becomes more dead, is raised in pitch, and the ring of the flask is lost.

The difference in character and in acoustic conditions between *tympanitic* and *amphoric* resonance on percussion may be learned by the simple experiment pointed out by me in the *Medical Times and Gazette*.

"Tympanitic resonance requires that the cavity percussed shall be full of air, but shall not communicate freely with external air.

"Amphoric resonance requires that the cavity percussed shall communicate freely with external air.

"That is, in the former case, the volume of air must be more or less confined; in the latter it must not be confined at all. The simplest demonstration of this difference may be made by procuring one of the vulcanized rubber balls, now common as toys. Seal up the small hole in it and percussion will yield *tympanitic* resonance; unseal the hole, and the resonance will still be imperfectly tympanitic, the aperture being too small for free communication with the external air; cut the hole large enough that no hissing is produced by the escaping air when the ball is suddenly compressed, *i. e.*, large enough to admit of free communication with the external air, and percussion will elicit *amphoric* resonance."

The varieties of percussion sound yielded by different bodies may be learned with great advantage by percussing a waterproof cushion filled with different kinds of matter, as water, air, wool, saw-dust. We may particularly observe the change in the percussion note while the cushion is more and more distended with air. Percussion should in fact be practiced upon bodies of all kinds, until the ear is perfectly familiar with the differences of each, and can identify them with the eyes shut.

The sound called *bruit de pot fêlé* may be very closely imitated by taking one of the vulcanized rubber balls with a small aperture in one side, already mentioned, and while it is placed upon a table, lay one finger upon it as a pleximeter, and then percuss the ball, the air from which puffs out at the hole: if a small glass or metal tube is pushed into the hole so as to project a little into the cavity of the ball, the sound of the air in escaping acquires a more metallic quality.

In this experiment we learn the acoustic conditions essential to the sound in question, viz., a space bounded by elastic compressible walls, containing air, communicating with the external atmosphere by a small orifice; it also teaches the method of percussion necessary to elicit the sound from the chest when these conditions exist, viz., "to give the impulse slowly and heavily, and allow the fingers to press forcibly on the part for some moments after it has been given."—*Walshe*.

Percussion of a vulcanized rubber foot-ball yields an admirable specimen of *metallic resonance*, and may well be added to the list of bodies upon which to practice percussion, as it also shows how a metallic sound may be yielded by bodies, under favorable circumstances, without their possessing the usual physical properties of metals. The sense of touch, as well as that of hearing, should be practised during these experiments. The detection of fluctuation by tact may be perfectly learned on the water cushion, and the elastic resistance of confined air on the vulcanized foot-ball. All varieties of *friction-sound* may be imitated by rubbing together beneath the stethoscope different materials, such as cloth, silk, velvet, leather, both wet and dry.

Not less important knowledge is an acquaintance with the influence of conducting media upon sound, and this may be easily acquired by listening to the tick of a watch, with different materials intervening between it and the stethoscope, or by producing rhonchus or crepitations in the elastic tube before spoken of, and listening to them in the same manner.

It would be a great mistake to suppose that these sounds, produced by experiments on inanimate bodies, are exactly like those audible in the chest; that is not pretended for a moment, although many of them are extremely similar, and convey an impression much more correct than any verbal description. The special advantage to be gained from the practice of these experiments consists in the education thus given to the ear; the acoustic conditions are in each case plainly seen; the sound can be listened to with ease, as often as requisite, free from the disturbances of the bedside; and hence, with a little perseverance, the susceptibility of the ear to distinguish and identify sounds and their causes will become so acute that the learner will be able to assign most sounds to their essential acoustic conditions, and will readily understand what he hears when he auscults or percusses the human chest.

We must not forget to guard against the danger of concentrating the attention on the special senses to the exclusion of that *common sense*, without which no learning, talents, or skill, and no accumulation of evidence, will insure wisdom in our judgments.

In an examination of the chest for the purpose of diagnosis, common sense dictates that no opinion should be formed of the importance or meaning of any delicate modification of sound until a comparison has been instituted between corresponding portions of the two sides of the thorax: and that we should not commence by seeking for signs of any *special pathological state*, which would be to begin the inquiry at what should be its end, and must terminate in confusion and loss of time. If we remember that it is matter with which we are concerned, and that the principal physical changes of which matter is susceptible are few and simple, we shall very much curtail the examination and secure a safe basis for each step as we proceed.

Having gained the best general ideas of the nature and seat of disease which inspection, palpation, and mensuration can afford, the first question in our future inquiry will be whether any part of the chest deviates from the percussion, respiration, or voice sounds of health. If not, the examination of physical signs naturally concludes. If, on the other hand, some deviation is detected, the question becomes, what physical change has occurred to account for it. This question will be most easily and quickly answered by dividing it as follows:

Is there consolidation (or increased density)?

Is there liquefaction?

Is there excavation?

Is there roughening of surfaces that should move smoothly upon each other?

Are these surfaces removed from normal apposition?

One or all of these divisions of the question may be answered either negatively or positively. And then succeed further questions, as to the extent, amount, etc., of each of the physical changes, and as to the relations existing between any two or more of them,—questions in which may be involved all the acoustic knowledge discussed in the preceding pages.

After this manner the inquiry may be kept within definite limits, and the mind prevented from wandering into desultory observations leading to no practical result.

Having arrived, as accurately as possible, at the physical properties of the altered structure, the question as to its pathological nature may be proceeded with at once. This will require the assistance of collateral information from other sources than physical signs, and the correct answer will be most easily arrived at by a process of negative argumentation, or, as it is called by logical writers, "abscissio infiniti," that is, by excluding, one by one, suppositions as to the nature of the change, and thus bringing the inquiry within such narrow limits that it is easily decided.

This important lesson, however, cannot be too constantly borne in mind—that *inspection, mensuration, palpation, percussion* and *auscultation* are means by which most valuable information may be gained of the conditions of organs within the chest; but that, either singly or combined, they are limited in their power to the revelation of certain physical conditions, which require the assistance of other knowledge than any of these means can afford, to give to each condition its true pathological meaning. The wise and skillful physician must have educated not only the ear, but every organ by which he can take cognizance of external things; his intellect must have been prepared to receive their impressions truthfully, to test their value, to arrange them, and to reason upon them; he must have practised the judgment in estimating the weight of evidence, restricting the flights of fancy, and in seizing with promptitude upon the just conclusion. But it should never be forgotten that the greatest wisdom may be shown by not coming to any conclusion at all on insufficient evidence.

The evidence of physical signs alone is often insufficient to form the basis of a safe diagnosis: kept to its proper place, it cannot be over estimated, but it must not assume to put aside other sources of information, upon which the physical signs depend for their interpretation. Thus, an assemblage of signs may be detected plainly indicating increased density of the structures within the chest, but this change in the physical condition may be due to a number of very different pathological states, and he who should conclude that dullness on percussion, bronchial breathing, and bronchophony, indicated tuberculous consolidation of the lung, might as easily be wrong as right; whereas, by bringing together the physical signs, and the results of clinical experience in all the other features of disease, those physical signs will each acquire a special force and meaning, and may then form the basis of a safe and accurate diagnosis.

II. THE DIAGNOSIS OF EARLY PHTHISIS.

THE physical signs of most serious diseases of the lungs and pleura are sufficiently marked and different to form a medium through which their changes of structure may be identified by the senses. Pneumonia, bronchitis, pleuritis, empyema, pneumothorax, have each their characteristic signs, so also, have the stages of consolidation, softening, and excavation, in tubercular disease. In its earlier progress, however, we cannot turn so satisfactorily to the results of physical examination. It is true that by this means disease may be detected in a comparatively recent state, that any *considerable* conglomeration of tubercular matter lying within acoustic reach of the surface of a lung can hardly escape detection, and that a *very abundant* deposit of isolated tubercles will, in most cases, interfere with the respiratory or other sounds of health in the chest sufficiently to give a strong suspicion of their existence. Of this, fig. 1 may be taken for an example: the miliary tubercles are extensively and abundantly scattered throughout the apex of one lung, and they did not entirely escape detection during life. But even in this case the physical signs enumerated are of a very questionable nature—neither singly nor collectively constituting an unmistakable proof of the existence of disease. In examining such cases in private practice—impressed with their deep social import—foreseeing the shadow that will be cast over the life of the patient, the gloom of apprehensive anxiety over that of his friends, if the judgment is adverse, and, on the other hand, the bright hopes that will be reinstated if it is favorable—the physician's heart may well sink despondently within him, when he reviews the evidence from which that signal judgment must be formed.

Again and again he may hesitate to frame it; again and again he may examine and listen, in the hope of discovering some unmistakable, some palpable signs of health or of disease; but to no purpose, for they are not there; and from a number of half-shadowed, spectral evidences, so slight and changing that their impressions can scarcely be retained while they are assembled, he must judge of the prospects of life or death.

This absence of reliable signs of the earliest stage of tubercular deposit cannot be too forcibly impressed upon the young practitioner, who, with creditable zeal, is too apt to think, and naturally prone to hope, that by sufficient diligence, experience, and care, he may insure that no tubercle shall escape his searching examination. In this belief he will be often led to fancy that he has detected the presence of tubercle where it does not exist, and to assume its absence while it really lies concealed.

This is the great disappointment which every man has to encounter who studies and practises physical diagnosis. That upon which he has set his heart—to detect the first shadow of consumption, scarcely yet resting on a life, before it is too late to drive it back—is the very dream which experience will most surely dissipate.

If physical diagnosis could detect consumption as soon as the first few spots of tuberculous matter were deposited in the lung, with the same certainty that it detects pneumonia or a cavity, we might well be content to sacrifice for this all that it could do besides. But, that it cannot, in its present state, accomplish this, and that there is no good reason to suppose that it ever will accomplish it, need surprise no one who will think calmly on the subject. It is not probable that the physical properties or the functions of a portion of lung-tissue should be sufficiently affected by a few scattered gelatiniform granules, to produce any recognizable signs of their existence. There may not be a dozen tubercles deposited in the whole lung; nay, more, there may not be one single microscopic speck deposited; and yet the disease may be working its stealthy inroads on vitality. We shall not, then, place implicit confidence in the physical diagnosis of early phthisis, but—while not neglecting this, while seeking from it all the aid it can give—we shall exert our most acute observation of symptoms, and diligently search into the histories of cases, in the hope that we may thus encounter some herald of the coming foe.

The following is an enumeration of the more delicate physical signs, by one or more of which it is presumed by different auscultators that the deposit of tubercle may be first indicated, before the occurrence of altered resonance on percussion:

1. Prolonged expiratory sound (Jackson and others).

"I am still of opinion that an increased expiratory murmur, provided it has not a bronchial or any other character than that proper to it, indicates nothing more than this—that the air, in passing out of the lungs, meets with some obstruction in the bronchial tubes."—*Skoda*.

2. Jerking respiratory sound.

"If the other causes of jerking rhythm can be excluded, which may or may not be difficult, this condition of rhythm, when limited to one apex, becomes a really important sign of tuberculization."—*Walshe*.

3. Deficiency in the respiratory sound.

"At the apex of one lung, coexistent with puerile vesicular murmur at the apex of the other, this is at all times to be considered a suspicious condition."—*Davies.*

4. Rough or harsh respiratory sound.

"Solitary tubercles, however abundant, do not necessarily interfere with the vesicular respiration."—*Skoda.*

5. Persistent signs of bronchitis, confined to one apex.

6. Cogged-wheel rhythm of respiratory sound.

"In some cases of incipient tuberculization, the tidal air seems to struggle against minute obstructions in the finer tubes, whence a rhythm of sound resembling that of a cogged wheel in rotation."—*Walshe.*

7. "Crumpling, buzzing, humming, kettle-singing, and arrow-root-powder sounds."—*Scott Alison.*

8. A variety of circumstances, the essence of which is comprised in the general terms "lessened respiratory action," with the addition of "wavy or jerking respiration and prolonged expiration."—*Edward Smith.*

III. THE VALUE OF CAVERNOUS SOUNDS.

CAVERNOUS respiration, voice and cough are sounds characteristic of a space of *considerable size*, with walls sufficiently hard and smooth to reflect sound —of such a space in fact as the tracheal tube. These very important sounds accompany and may be present in larger dilatations, provided the cavities are not too much filled with their fluid contents, or cut off from free communication with pervious air tubes, and if the walls are not too soft to reflect sound. These cavernous or tracheal sounds derive their value, in a diagnostic sense, from the following circumstances:

1. There is no part of the chest-wall, after passing the first bone of the sternum, from which they can be detected, in a state of perfect health—that is, if auscultation is properly performed; for in listening above the clavicles, if the stethoscope is carelessly directed towards the trachea, instead of in a line with the perpendicular axis of the chest, tracheal sounds may be heard.

2. Although the sounds of bronchial tubes may become audible at the surface of the chest, through increased conducting power of the structures inter-

posed, these tubes are not large enough in health, except perhaps at their bifurcation, to give sounds, like those from the trachea.

3. Cavernous sounds are produced under conditions which almost insure their reaching the thoracic parietes: viz., (*a*) a cavity so large that the vibrations of a considerable volume of air may be transmitted to its walls; (*b*) these walls must present a large surface from which vibrations may be conducted to the ear; (*c*) the walls must be of a certain density, which implies, almost of necessity, increased density of the surrounding structures—that is, increased power of conducting, sound; (*d*) a cavity so large and so defined, that it will almost necessarily, communicate freely with some of the bronchial passages upon which it has intruded, thus acquiring free ingress and egress of air during respiration, speaking, or coughing.

Hence the sounds which in health are audible only on ausculting the trachea, when heard at any part of the chest except over the trachea and its bifurcation, positively indicate the existence, not only of a cavity, but of a cavity larger than is consistent with health; and thus they have acquired the name *par excellence* of "cavernous"—as distinguished from "bronchial," from which they are separated by difference of size. Nevertheless, we find cavities, the result of disease, large enough to be of serious consequence, and yet no cavernous sounds are produced by them. Therefore, we learn that, although these sounds are positive evidence of the existence of a cavity, their absence is not positive evidence that no cavity exists; they thus lose much of their value; for in truth it is just in those cases in which they are not audible, that it is most important to detect the existence of disease—cases in which the general symptoms are not very marked, and which are not too far advanced to afford a reasonable prospect of recovery.

IV. THE IMPORT OF HÆMOPTYSIS AS A SYMPTOM.

IN a paper published in the Transactions of the Royal Medical and Chirurgical Society, Dr. Dobell gives an analysis of 100 male cases of hæmoptysis.

The cases were sifted in the following manner:—1. All cases were rejected in which the blood had never been seen in any other form than "*streaks in the phlegm.*" 2. All cases were rejected unless the heaviest weight before the occurrence of hæmoptysis could be stated from *actual weighing*, and reasonable evidence given as to whether this had been the average weight up to the time of first hæmoptysis. 3. All cases were rejected who could not stand a searching cross-examination as to the time at which the *first loss of weight*, if any, had begun. 4. It was soon found necessary to reject all females, for the following reasons:—*a.* Their weight could not be relied on. *b.* Pregnancy, lactation, etc., were constant sources of fallacy. *c.* Hæmoptysis was found to be complicated with climacteric and other derangements of the menstrual functions. *d.* It was difficult to obtain any reliable facts. 5. All cases were rejected in which there was reasonable suspicion of cardiac complications. 6. All cases were rejected who could not give a fairly succinct account of the onset of the cough. 7. All cases were rejected who could not give an approximate estimate of the quantity and character of the expectorated blood in the first and subsequent hæmoptysis. 8. And, finally, after the inquiry had been completed, all cases were rejected if it was found, on comparing the principal statements, that they were inconsistent with one another.

The elements of these rigorous rejections are such that they do not give the cases a selected character in the sense of invalidating their claim to represent an *unprejudiced average* of cases of hæmoptysis occurring at a public hospital. They may, therefore, be considered to form a fairer basis for statistics than if no selection had been made, having the great advantage that incomplete and unreliable reports are excluded.

The analysis of these cases, the results of which are epitomized in an abridged table published in the Transactions of the Society, leads the author to the following conclusions.

Hæmoptysis as a symptom may be thus classified:

1. In a large number of cases it is simply the result of congestion and disintegration of a highly vascular organ in the course of a disease of constitutional origin.

2. In a large number of cases it is simply the result of congestion and disintegration of a highly vascular organ in the course of diseases of local origin.

3. In a certain number of cases it is simply the result of accidents temporarily over-distending the vascular system of the lungs, and leading to rupture in the same way as similar over-distension leads to rupture of vessels in other parts of the body. Whether such over-distension is competent to cause rupture of vessels, the walls of which are not previously diseased, is a very wide question.

4. In a certain number of cases it is the result of the bursting of small aneurisms in the lungs, formed in the course of lung disease.

As a cause of lung disease and constitutional decline, hæmoptysis is considered to be one item, and that a very occasoinal one, in a large and important group, embracing *all foreign substances which find their way into the peri-vascular and peri-alveolar tissue* of the lungs, and by their irritation there, set up lymphatic (adenoid) and connective tissue cell proliferation and its consequences. Of this important group the following are some of the principal constituents:—*a*. The dust of flint, coal, iron, and other substances inhaled by workers in different dusty trades. *b*. The products of inflammatory destruction of tissue. *c*. The products of catarrhal affections. *d*. The débris of blood, and of tissues disintegrated by the extravasation of blood. *e.* Albuminoid tissue disintegrated by peroxidation in true tuberculosis (Dr. Dobell's hypothesis). *f.* Accumulation of blood débris in the alveoli.

The disintegrated albuminoid tissue is the irritant which sets up that hyperplasia of adenoid tissue and its results, so well described by Portal, Virchow, Sanderson and Rindfleisch. But whereas they place this hyperplasia first among the pathological changes of tuberculosis, precedence should be given to the peroxidation and disintegration of albuminoid tissue, of which the hyperplastic changes are but the effects, the order of events being:—*a.* Deficiency of fat in the blood. *b.* Peroxidation of albuminoid tissue. *c.* The production of disintegrated albuminoid tissue, the result of peroxidation. *d.* Hyperplasia of adenoid tissue, the result of irritation of the absorbent system engaged in removing the disintegrated tissue.

Whether the disintegrated albuminoid tissue or the resulting diseased adenoid tissue shall be called "Tubercle" Dr. Dobell thinks of little consequence, so that the distinction in the order of events is borne in mind.

V. ON WINTER COUGHS, AND THE RELATIONS OF BRONCHITIS AND EMPHYSEMA.*

Introductory Remarks—Importance of Winter Cough—Necessity for a searching enquiry into the History of Cases—Arrangement of Cases into Five Clinical Groups—Questions as to the relation between Emphysema, Bronchitis, and Winter Cough.—Mode of production of Emphysema discussed.

HOWEVER much a medical man's tastes may tempt him to devote his time and energies to the purely scientific departments of his profession, he must never lose sight of the fact that he is neglecting his duty in proportion as he allows himself to be led away from such studies as have a practical bearing upon the prevention, relief, and cure of disease. There ought to be no alternative in his mind, between giving up practice altogether, and devoting all his best energies to making his practice beneficial to his patients in the highest possible degree. I make these remarks in explanation of my having, in these lectures, passed by many points exceedingly interesting as subjects of sci-

---

*This and the remaining chapters of this part are from Dr. Dobell's lectures, delivered at the Royal Hospital for Diseases of the Chest.

entific speculation and enquiry, in order that I might devote all the time at our disposal to those which come more strictly within the limits of practical medicine.

Again, I can but think that, as practitioners of medicine, we ought to look with far more interest upon anything that conduces to the cure or relief of the diseases which affect the largest number of persons, than upon any rare and solitary cases, however, curious they may be. Therefore, I have not hesitated to devote a large amount of time and study to the very common and well-known classes of cases which form subject of these lectures.

I have included them all under the crude name of "*Winter Cough*," because it expresses the one conspicuous symptom, common to them all, which especially brings such cases under the eye of the physician. All the patients had a cough, which was either limited to the winter season, or was much aggravated during that part of the year.

However tedious and wanting in the excitement of novelty a common case of Winter Cough may be to the medical practitioner, there are but few complaints which so painfully absorb the interest and attention of the patient; and as such cases are extremely numerous in all ranks of society in this climate, they represent an enormous amount of human suffering, and from this fact alone demand our most earnest consideration.

I need hardly tell you that at this hospital such cases abound in every form and variety, and afford the widest possible field for enquiry, whether it be into their symptoms and physical signs, their course and treatment, their consequences and terminations, or their causes and natural history.

As two or more winters usually pass before the tendency of the complaint to recur or to become habitual is established in a patient's mind, these cases of Winter Cough have always a history more or less long, and of which it is often not very easy to get a correct account. But in this history lie just those points of the case which are essential to a proper understanding of the causation of the disease, and of the prospects of cure or relief. It is into this history, therefore, that we should always push our enquiries with great perseverance, taking the utmost care not to be misled by the erroneous and conflicting statements which patients are sure to make, unless we give them time to think over the past before committing themselves to an account of it.

In order to guard as much as possible against this source of fallacy, and at the same time to secure uniformity in a large number of reports, so as to admit of their chief points being tabulated, I have been accustomed to give to hospital patients a printed list of questions to think over at their leisure, before committing themselves to the answers.

These questions are forty-one in number and refer to the short breath, the cough, the taking of colds, the past illnesses, the occupation, the dwelling, the habits, and the family history of the patient. They are as follows:

FORM OF ENQUIRY INTO THE HISTORY OF WINTER COUGH.

BREATHING.	When did the breath first begin to be short on going up stairs or hills?
	In what other way were you ill when the breath first began to be short?
	What sort of health had you before the breath began to be short?
	Since your breathing first became short, has it ever been otherwise than short? If so, when?
	What do you think most inclined to make your breathing short?
COUGH.	When did you first have an attack of cough?
	What sort of attack was it?
	What else was the matter at the time?
	How often and when have such attacks returned?
	How has the breath been between the attacks of cough?
	Have you ever been quite free from cough since the first attack? If so, when?
	When you catch a cold, does it affect first the chest, the throat, or the nose?
	Describe the symptoms of the attacks of cold which leave a cough?
	What gives you cold most easily?
	What gives you cold most often?
	Is your cough much worse at times from any other causes than fresh cold?

Diagnosis of Pulmonary Diseases. 53

PAST HISTORY, &c.
- What illnesses have you had within memory not already stated?
- Do you attribute your complaint in the chest to either of those illnesses?
- If you do, to which do you attribute it, and what reason have you for doing so?
- If your mother is living, ask her these questions, and state what she thinks.
- If your mother is living, ask her whether you were a strong and healthy, or a weak and delicate child and give her answer.

FAMILY HISTORY.

MOTHER.—If living, what age and what health?
    If dead, what age at death, and the cause of death?
    What health had she during life?

FATHER.—If living, what age and what health?
    If dead, what age at death, and the cause of death?
    What health had he during life?

BROTHERS.—How many living?
    What are their ages, and what health have they?
    How many dead?
    What were their ages, and the causes of death?
    What health had they in life?

SISTERS.—How many living?
    What are their ages, and what health have they?
    How many dead?
    What were their ages, and the causes of death?
    What health had they in life?

OCCUPATION.
- What is your present occupation, and how long have you followed it?
- What other occupation have you followed, and at what periods?
- What are your hours for business and for meals?
- In what sort of place do you live by day?
- In what sort of place do you live by night?

DIET, HABITS, &C.	Do you take meat, and vegetables, and bread?
	Have you always done so?
	What fermented liquors do you drink, and what quantity per day?
	Have you always taken the same?
	Do you live regularly, and are your spirits usually good or bad?
	Do you smoke?
	How long have you done so?
	Has it had any effect on your complaint?
	Has anything you do any effect on your complaint, and if so, what?

In collating the notes of a large number of cases the histories of which are taken in this manner, and to which the physical signs and symptoms are attached, I have found that they may be very simply arranged in five clinical groups.

1. Cases in where there are physical signs of Emphysema and not of Bronchitis, and in which there is no history of previous Bronchitis.

2. Cases in which there are physical signs of Emphysema and not of Bronchitis, but in which there is a history of previous Bronchitis.

3. Cases in which there are physical signs of Bronchitis and not of Emphysema.

4. Cases in which there are physical signs both of Emphysema and of Bronchitis.

5. Exceptional cases in which there are no physical signs, either of Bronchitis or of Emphysema.

You will at once observe that these groups give us Emphysema and Bronchitis as the two conditions of disease which, with the exception of a few cases, are ever present, either singly or combined, when there is Winter Cough.

It is evidently, therefore, the leading point of interest, in a practical sense, to know what is the relation of these states—Emphysema and Bronchitis—to each other and to the complaint in question, viz., Winter Cough.

Is the cough produced by Emphysema, by Bronchitis, or by both, or is it dependent upon some other condition which accompanies both the Bronchitis and the Emphysema?

Is the Bronchitis produced by Emphysema, or Emphysema by Bronchitis?

Is the Bronchitis or the Emphysema produced by the Cough?

In what way, if any, do Bronchitis and Emphysema influence each other?

These are questions which lie at the bottom of our treatment of the disease, whether it be preventive, curative, or alleviative; and I am sorry to say that they cannot all be answered so simply as might at first sight appear, for some of them have already engaged the best attention of excellent pathologists and practical physicians, and the conclusions they have respectively come to have been very different.

As my own views on the subject are in opposition to those of some of the physicians for whose opinions both the profession generally and myself entertain the greatest respect, I cannot put them forward without to some extent attempting to support them by arguments, and I consider that a right understanding of this part of the subject is so essential to the whole question of the treatment of Winter Cough that I cannot pass it by. I will, however, attempt to put these views before you as as briefly as a due respect for the opinions of others will permit.

The great question of dispute relates to the mode of production of Emphysema.

How is Emphysema produced?

1. Is it due to the forcible expansion of the air vesicles of the lungs during the inspiratory act?

2. Is it due to a compensatory dilatation of the air vesicles, rendered necessary by the collapse of neighboring portions of lung?

3. Is it due to a degeneration of the tissue of the air vesicles, which renders them incompetent to withstand the normal dilating influences during normal respiration?

4. Is it due to the forcible expansion of the air vesicles of the lung during the expiratory act?

You are aware that there are powerful and accomplished advocates for each of these four propositions, and when I assert, as my strong conviction, that the fourth alone can be maintained*—that the production of Emphysema can be more satisfactorily explained by the expansion of the air vesicles during expiration, than by any other cause, and that this belief is essential to the proper treatment of Winter Cough—I feel that I am bound in some measure to support the opinion I hold by arguments which may commend it to your respect, although I cannot venture to occupy your time with such a detailed and abstruse discussion as would be necessary to answer all objections urged on the various sides of the question.

The first thing to do is to show that the phenomena of Emphysema can be satisfactorily explained by the expiratory theory of its production.

The second is to show that the conditions required by this theory for the production of these phenomena are supplied by disease.

And the third is to show that evidences of the existence of these conditions are to be found in the clinical histories and features of cases of Emphysema.

Having done these things, I will briefly point out some of the vital objections to the other theories to which I have referred.

---

* I allow collapse of lung and degeneration of tissue a place among the conditions *predisposing* to Emphysema in certain exceptional cases, but I maintain that the over-distension of the cells is produced by the expiratory act.

Emphysema of the lung, as you well know, may be limited to a few air-cells, or may affect one or more lobules, a whole lobe, and even the whole of one or both lungs. It essentially consists in a dilatation of the air-cells, which may be slight or extreme, and in attenuation and rupture of the dilated walls of the cells, so that contiguous cells are made to communicate with each other; and in the further progress of the disease the partitions between the cells almost entirely disappear, and large irregular air sacs are formed by the coalescence of neighboring cells. In this way the bulk of the affected lung is increased, and its normal elasticity is lost.

You are probably aware that a state resembling Emphysema may be produced in a healthy lung after death by forcibly blowing into the air tubes, and thus over-distending the vesicular structure. Dr. John Hutchinson found that in pumping air after death into the human chest a pressure of nearly 12 ozs. avoirdupois upon every square inch of surface was sufficient to rupture the pulmonary substance. There can be no doubt that during life the air-cells will admit of much greater distension before rupture of their walls occurs than after death, especially if the distending force is applied very gradually. But in essential characters the distension and rupture which occurs when the lungs are forcibly distended after death are analogous to that which occurs during life in the production of Emphysema.

I am aware that this experiment has been supposed to favor the inspiratory theory of the production of Emphysema, but a moment's thought will teach us that it is clearly no illustration of an inspiratory operation. Inspiration is distinctly a suction force exerted from without, in which the air simply follows the expansion of the chest-walls, and of the lung, and is accompanied by no *vis a tergo;* whereas, in the experiment quoted, the lung is distended entirely by means of a force exerted from behind.

We see, then, from this experiment, that all that is necessary for the production of Emphysema is an undue pressure of air upon the internal surface of the air-cells.

You are no doubt familliar with the fact that what is called "Interlobular Emphysema" may be produced by forcible expiratory acts. That during such acts one or more air-cells may become so over-distended, that rupture takes place, and the air from the lung is poured into the cellular tissue between the lobules, whence it may find its way into the cellular tissue of the mediastinum, and thence into that of the chest and neck. I well remember a case of this kind, which was brought to my notice by Dr. Niell, of Aldersgate street. He was attending a lady in a difficult confinement, and suddenly, during a violent expulsive effort, the patient's neck and shoulders swelled up to an enormous size, and the cellular tissue was found to be full of air. Many such cases have been witnessed, always occurring during some violent respiratory effort, as during parturition, defæcation, lifting heavy weights, or coughing. A number of cases, in which this accident occurred during fits of coughing, are recorded by M. Guillot, (Archives générales de Médecine, vol. ii., 1853), and many others may be found scattered through other works and periodicals.

From these two facts, then,—(1) that distention and rupture of the air-cells, similar to Emphysema, have been produced experimentally by undue pressure of air, applied to the inner surface of the cells, and (2) that similar distention and rupture of the air-cells have been produced accidentally by forcible acts of expiration,—it may be concluded that vesicular Emphysema may be produced by pressure applied to the inner surface of the air-cells during expiration.

Now, although by those who only approve of complicated modes of proof in scientific matters, this may appear a somewhat short and off-hand way of deciding a diffiicult question, I think you will find it unanswerable, and that is the important point to look to. If we have once seen unmistakably that a thing can be done, and has been done, we need not trouble ourselves much with arguments which attempt to prove theoretically that it cannot be done. And here we have two very simple illustrations, the one showing that it can be done, and the other that it has been done; and therefore, to my mind, the question is so far settled.

The next portions of the proof required of us, viz., to show that the conditions necessary to the expiratory theory are supplied by disease, and that evidences of the existence of these conditions are to be discovered in the clinical histories and features of cases of Emphysema, cannot be disposed of in so short and off-hand a way. It is beset with all those difficulties which must ever surround questions which deal with complicated vital acts conducted within the body; and in this case an additional difficulty and a fertile source of error is introduced by the circumstance that considerable periods of time are in most cases occupied in the development of the effects we have to deal with, and by the circumstance that the cause may have ceased to exist at the time the effects are brought under our observation. We have, therefore, to search in one individual for the cause of effects which we witness in another: a mode of inquiry than which nothing can be more puzzling and open to fallacy. And when the effect is discovered and the cause found wanting, it is at once competent, for those who believe that the assigned cause is not the true one, to bring these cases forward as proofs of the correctness of their opinions. Our great endeavor, therefore, should be to find a mode of observation which shall as far as possible remove this source of fallacy.

It has appeared to me that this source of fallacy must remain so long as the post-mortem examination of the patient is depended upon for the explanation of the phenomena of his disease. I am speaking especially of the disease now under our consideration, although the remark may apply to others. A simple illustration of what I mean is this:—The examination of the lungs of a man who has long suffered from attacks of Spasmodic Asthma may show well developed Emphysema. But not a trace may remain of the bronchial spasm, the recurrence of which produced the Emphysema; whereas, on the other·hand, there may be found many other changes in his bronchial tubes, or in the parenchyma of the lung, due to causes totally unconnected with either the Asthmatic Spasm, or the Emphysematous air-cells, while the history of the case may be totally incompetent to determine whether or not any relation existed during life between the pathological changes discovered after death. On the other hand, there were, in all probability, periods in the course of the case when symptoms and physical signs, properly interpreted, might have told what morbid causes were in action, and what organic changes were being produced by them, the course of both being watched and traced.

In making these remarks, I wish particularly not to be misunderstood. I do not in the least undervalue the great importance of morbid anatomy, when the conclusions drawn from it are kept within legitimate limits.

In the next place, I have to show what are the physical conditions required in the chest to produce Emphysema by expiratory acts, and what clinical means we possess of ascertaining that such conditions exist in any given case.

In the operation which we commonly call "straining," we first take a deep inspiration, by which the thorax is distended and the lungs filled with air, then close the glottis, so as to keep the air locked into the chest, and with the thorax thus tightly distended, we put the abdominal and other expiratory muscles into forcible contraction.

If you were to inflate a small bladder, and then grasp it forcibly in the hand, you would not be much surprised if the bladder were to burst at any point where its walls happened to be weakest and least supported by the hand.

We cannot, then, be surprised to find that when the distended lungs and chest-walls are forcibly pressed on all sides by powerful muscles, while the escape of the air by the glottis is prevented, they occasionally give way at some of the weakest points. This happens occasionally to the chest-walls themselves, so that portions of lung are actually protruded through them, constituting thoracic herniæ. But it more commonly happens that the delicate air vesicles burst at some part of the lung, where the external pressure happens for the moment to be the least, and give rise to inter-lobular Emphysema.

It is, then, very clear, that during these efforts of straining; when, after inspiration, an obstacle is put in the way of expiration, and at the same time muscular pressure is brought to bear upon the outside of the chest, air is driven with force against the inner walls of the air-cells; that, in fact, so far as the air-cells are concerned, the operation is similar to the experiment of blowing into the lungs down the bronchial tubes. And, as we might reasonably expect, the effects of two similar causes are alike, viz., the distension and rupture of the air-cells.

These, then, are the physical conditions required in the chest to produce a strain upon the inner surface of the air-cells during expiration. But it remains to show that these conditions are supplied by the circumstances which precede and accompany the occurrence of Emphysema, and also that the external pressure upon the lungs, exerted during these periods of straining, is subject to be weaker in some places than in others.

It is singular that it should not have been recognized, simple and self-evident as it appears, that what applies to the interference with the calibre of the small air tubes applies equally to the large ones. If the calibre of a tube, be it large or small, is justly proportioned to the transmission of a certain volume of air and no more in a given time, any decrease in the calibre must necessitate either a longer period of time for the passage of the same volume of air, or an increase in the rate at which the air passes. That is to say, a larger slower tide in the first case, and a smaller faster tide in the second case, must pass in the same period of time. But it is clear that a greater pressure from behind is required to urge the smaller-faster tide than that required for the larger-slower tide, if equal volumes are to pass in the same time, as stated, and in proportion to this increase of pressure there will be increase of friction upon the walls of the tubes.

It is evident, therefore, that, as the large bronchi, the trachea, larynx, and nasal passages, all are nicely adjusted to carry a given volume of air in a given time, any thickening of their linings, contraction of their walls, or any other cause which diminishes their calibre, must interfere with this adjustment, and necessitate either a slower tide through them and thus an increase in the time occupied by the passage of a certain volume of air, or an increased *vis a tergo* and a quicker tide through them, maintaining the just period at which the given volume passes, but doing so at the cost of increased pressure and increased friction.

In either of these cases the normal respiratory act is seriously disturbed.

It is true that in the case of the nasal passages, so subject to temporary obstructions, special provision has been made, in the power of breathing through the mouth, to avoid, to some extent, the interference with respiration which might otherwise so frequently occur. But below the fauces there is no such safety valve, and the changes in calibre below this point must of necessity disturb the respiratory adjustment.

For these reasons I consider that any decrease of calibre in the nasal passages, larynx, trachea, or large bronchi, must be considered as obstructions to the expiratory tide, and that they are, in fact, much more important obstructions as regards the production of backward pressure than any that can occur in the small air passages, because they interfere with the main thoroughfares of the lungs, whereas the others merely stop the smaller ramifications through which there is not only much less traffic, but the influence of its obstruction is limited to a small number of cells.

It follows, then, that I include among causes of *sudden* strain on the air-cells circumstances which have been heretofore overlooked, viz., the acts of sneezing, and of blowing the nose, when they are convulsively or violently performed, and when an abnormal obstruction is placed in the way of the outward tide, as in nasal polypus and in tumefaction of the naso-pulmonary mucous membrane.

And I include among causes of obstructed outward tide during *ordinary* respiration, circumstances which have heretofore been overlooked, viz., catarrhal thickening of the mucous lining of the nose, larynx, trachea, and large bronchi.

During all acts of coughing, a certain backward shock of air occurs before the glottis is opened, and before the body to be expelled can be ejected. But this becomes a cause of Emphysema when the cough is unusually convulsive and severe, and when some obstruction unusually difficult to remove is placed in the way of the outward tide.

It has not, so far as I can learn, been recognized, that when the bronchial tubes are diminished in calibre an obstruction is placed in the way of the outward tide, which does not affect the inward tide to the same extent. Yet such is undoubtedly the fact. It occurs in this way:—The act of inspiration consists in enlarging the capacity of the chest by muscular force, and thus removing from the whole lung the circumferential pressure of its own elastic force; the consequence is, that the expanding lung expands the enclosed bronchial tubes to the fullest extent, and thus favors the influx of air. The normal expiratory act consists in a simple recoil of the elastic lung, followed by that of the elastic chest-walls; a pressure is thus exerted on the circumference of the bronchial tubes, which tends to diminish their calibre, and favors the exit of air; but this expansion during inspiration, and compression during expiration, is all calculated for in the adjustment of normal respiration, and no backward pressure is produced; but so soon as a sufficient diminution has occurred in the calibre of the air passages to require a greater expiratory force than is supplied by the elastic recoil of the lung, an entirely different relation is set up between inspiration and expiration.

So soon as the additional force of *muscular expiratory efforts* is called for, the circumferential pressure upon the bronchi acts unfavorably to the outward tide; the air-cells are pressed upon the walls of the bronchi which they surround, and act as causes of increased obstruction to the already diminished passages. In this way diminished calibre of the air passages tells more unfavorably upon expiration than upon inspiration, and becomes a cause of backward pressure of air upon the cells of the lungs during the expiratory act, even in ordinary breathing.

The mode of operation may be watched in the nose, only that there it is exactly reversed. The tendency of the walls of the nose, not the alæ nasi, is to collapse during inspiration, and to expand by the pressure of air during expiration. If we experiment upon ourselves, while suffering from the tumid state of nasal catarrh, we shall find that it is often impossible to inspire through the nose—the suction-force exerted by the chest producing collapse of the nasal walls and complete obstruction—when it is comparatively easy to expire through the nose in consequence of the passages being opened by the outward tide of air. On the other hand, if we try a similar experiment with the chest during the tumid stage of bronchial catarrh, we shall find that it is easy to inspire, when expiration is attended with labor in consequence of the outward pressure upon the obstructed tubes still further diminishing their calibre.

Supposing it to be granted that, under the circumstances which we have considered, a backward pressure does occur upon the inner surface of the air-cells, it remains to be shown that the external pressure upon the lungs during these periods of strain is subject to inequalities.

The principal objection raised to the expiratory

theory of the production of Emphysema is an assertion that such inequalities do not exist, an objection which is very well set forth by Dr. Gairdner in the following words:—"Even when the air vesicles are maintained at their maximum or normal state of fulness, by a closed glottis, any further distension of them is as much out of the question as would be the further distension of the bladder, blown up and tied at the neck, by hydrostatic or equalized pressure, applied to its entire external surface." (Monthly Journal of Med. Science, vol. xiii.) This is, in fact, the stand-point in the argument; and whatever force it might have as a proof that over-distension of the air-cells cannot be produced by expiration is turned in the opposite direction, if it can be shown that it breaks down at its most vital point when submitted to the test of experiment; if, in fact, Dr. Gairdner's "bladder blown up and tied at the neck," instead of having "equalized pressure applied to its entire external surface," has the pressure applied unequally, as if grasped by the hand. And whereas there is no evidence to prove the equalized character of the pressure, there is plenty in favor of the unequal pressure.

It seems to be assumed by those who use this argument, that in ordinary normal respiration the air is forced from the lungs by the elastic pressure of the thoracic walls. If this were the case it would certainly be necessary that the weakest parts of those walls should at least be strong enough to prevent any eccentric yielding of the parts to be compressed, viz., the lungs. But I believe this assumption to be entirely false. In a state of health the elastic contractile power of the lung itself is so much in excess of the power of the chest-wall to act upon it, that when the thorax has contracted as far as it is possible for it to go, the lung is ready to contract still further, and is actually held back from so doing by its connections with the chest-wall. This has been shown by Dr. Hyde Salter, who found in some very carefully performed experiments that the lung of a dog, when released from the thoracic parietes, undergoes a reduction of one-fifth of its volume, and that in the human subject the "elastic contractility of the lung is always drawing on the inner surface of the chest," so that, when the ribs have subsided to the exact point at which of themselves they would be disposed to stop, they are carried a little further, and only stop when the lungs have drawn them so far beyond their proper point of rest that the force of recoil thereby generated is exactly equal to the contractility of the lungs." (Lancet, July 29th, August 5th, 1865.) The consequence is, that the chest-wall, in ordinary expiration, has only to follow the contracting lung, so that it shall be in a position for expansion when the time arrives for the next inspiration. There was no need, therefore, that the singular equality of pressure, assumed by the argument referred to, should be provided; and accordingly, in the wise economy of nature, it has not been provided. But when the outward tide is interrupted by some abnormal obstruction and a call made upon the elastic recoil of the lung greater than that for which it is prepared, the lung, instead of taking the lead and being only followed by the chest-walls, falls back upon them for assistance, and they in their turn being incompetent by their elastic force to overcome the difficulty, fall back upon the expiratory muscles to assist in the act of compression. In this operation all the inequalities of pressure which may exist become opportunities for the over-distension of the air-cells, as they are urged in this direction and in that, in the attempts to overcome the obstruction to the outward tide of air.

A striking illustration of the way in which the lungs may be distended during forced expiration in any direction in which the outward pressure is deficient, was shown in the case of M. Groux, when exhibiting in this country at the different Medical Schools. During a violent expiratory act the lung of one side came forward in the upper part of the fissure which existed in his chest-walls, and formed there a distinct elongated resonant tumor, but no such result took place during inspiration. That, in the normal chest, inequalities of circumferential pressure exist during expiration, has been clearly shown by Sir Wm. Jenner and others.

Many other sources of inequality of pressure upon the lungs might be enumerated, amongst which I would suggest the mobility of the heart, which permits it to be pushed downwards, as proved by the change in its position which actually takes place when the upper parts of the lungs become seriously Emphysematous.

Again, the possibility of muscular contractions being more powerful and complete in some sets of muscles and in some sets of fibres than in others at the same time, is familiar to us all in other parts of the body; and it is only reasonable to suppose that similar irregularities occur when, from any cause, the muscles of expiration are called upon to perform more persistent duties than those for which they are normally intended.

A remarkable coincidence of backward pressure upon the air-cells and absence of uniform external compression, must occur when convulsive acts of coughing or sneezing are rendered abortive of their effects by the obstructing body, which those efforts were intended to remove, obstinately resisting their force; for the whole muscular force discharged upon the obstructing body must be thrown back upon the volume of air behind it at the very moment when the expiratory muscles have fallen into relaxation and disorder after their convulsive effort.

But even if we could not thus identify the modes and occasions of irregular and unequal pressure upon the superficies of the lungs, there are facts to show that such inequalities must exist.

The fact, for example, that in lobar Emphysema the surface of the lung is often marked with the impressions of the ribs, shows that the pressure from the chest-walls is not so exactly uniform but that the ribs press more and the intercostal spaces less. Again, in lobular Emphysema "the Emphysematous lobules are seen on the surface of the lung, protruding beyond the level of the surrounding tissue and along the margins of the lobes; they often form projections of considerable size, in some instances becoming developed into the so-called appendages." (Dr. Waters.)

If the equality of external pressure can so completely fail in one spot as to allow such projections to occur and to become permanent, it is clear that it may occur in other parts, and probably in different parts of the chest at different times.

But, in truth, we need hardly go further than the simple proposition with which I started—that distension and rupture of air vesicles producing interlobular Emphysema have again and again occurred during violent expiratory acts, is an unanswerable proof that inequalities of external pressure must occur during expiration sufficient to account for all the effects we require.

It has now been shown that distension and rupture of the air-cells may be produced experimentally by blowing forcibly into the lungs from the trachea, and that distension and rupture of the air-cells, causing interlobular Emphysema, have been produced by forcible expiratory acts. Presuming it to be granted from the arguments which have been adduced—

1. That obstruction to the outward tide of air may be produced by disease.

2. That such obstruction will cause pressure on the inner surface of the air-cells during expiration.

3. That pressure so caused may dilate the air-cells (as in Emphysema) if the pressure on the superficies of the lungs is subject to be unequal in force at different parts during one act of forced expiration.

4. That such inequalities of pressure do occur.

Then it follows that we have shown as proposed—

1. That the phenomena of Emphysema can be satisfactorily explained by the expiratory theory.

2. That the conditions required by this theory for the production of these phenomena may be supplied by disease.

It now remains to show that—

Evidences of the existence of the required conditions may be found in the clinical features and histories of cases of Emphysema.

Let us, then, enumerate the principal conditions for evidence of the present or past existence of which we have to search in cases of Emphysema.

The required conditions consist of any circumstances which may press the air in the lungs back upon the inner surface of the air-cells with greater force than their elastic properties are competent to resist, and thus deprive them of the power of again resisting a distending force which previously they were equal to withstand. They may be divided into two classes—

CLASS A.—Circumstances which may *at once* overstretch the air-cells.

CLASS B.—Circumstances which may gradually overstretch the air-cells.

These two classes of circumstances often succeed each other, and thus what one begins is carried forward by the other.

The conditions included under Class A, or such as may *at once* overstretch the air-cells, may be conveniently considered under two heads—

1. Violent expiratory acts, as in defæcation, parturition, lifting heavy weights, and the like, performed with a closed glottis.

2. Convulsive expiratory acts, as in whooping-cough, croup, layrngitis, fits of sneezing, nose-blowing

and the like, when of undue force and opposed by undue resistance.

The second class of circumstances, or such as may *gradually* overstretch the air-cells, may be conveniently under four heads—

1. Ordinary acts of coughing, when the free expulsion of the air in the chest is prevented by some unexpected obstruction at the moment when the glottis is opened and the expiratory paroxysm has culminated.

2. Ordinary acts of sneezing and nose-blowing, opposed by considerable obstruction of the nasal passages, and frequently repeated.

3. Ordinary respiration, when the outward tide is sufficiently obstructed by narrowed naso-pulmonary air passages to require muscular expiration.

4. Ordinary acts of coughing, sneezing, or nose-blowing, when some portions of the air-cells are deprived of their normal circumferential supports.

[The editor omits the cases detailed by the author to support the above propositions. It is sufficient to say they are carefully analyzed and to the point.]

VI. THE DIAGNOSIS OF NARROWED AIR PASSAGES.

I HAVE shown that narrowing of the naso-pulmonary air passages holds the most important place among causes of Emphysema; and I now come to consider the means which we have at our disposal, by which to ascertain, during life, whether or not the respiratory tubes are narrowed in any given case. I shall be able to point out some very simple physical signs, the importance of which does not appear to have occurred to other observers than myself, as they are not mentioned in books, but which are, I think, capable of satisfactory demonstration.

I refer especially to certain modifications of *pitch* in the inspiratory and expiratory sounds.

In order to give these their proper meaning and value, it is necessary to bear in mind a few acoustic details. I shall not trouble you with more than a very brief recital of the recognized rules with regard to pitch which are necessary to the present question.

Of all the qualities of sound, except loudness, pitch is that most easily and unmistakably distinguished. It refers, as you are aware, to the differences popularly known as "high" and "low;" differences which if they are at all wide are caught at once, even by the uneducated ear, and which to a good musical ear properly educated are distinguishable when reduced to a mere fraction.

For this reason I attach the greatest practical importance in physical diagnosis to all modifications of *pitch*.

Timbre is a most valuable attribute of sound, and when thoroughly understood is competent to give very important indications in disease; but it is far more subtle in its nature than pitch, and therefore more subject to be misunderstood. Its appreciation by the ear is less amenable to education, and its modifications are due to such a variety of occult causes that it is beset with sources of fallacy, and above all, its characters are exceedingly difficult to describe.

Pitch, on the other hand, is not only easily detected and easily described, but there is one constant condition upon which all its modifications depend, viz., the rate of vibration. All continued sound is but a repetition of impulses, and the pitch depends upon the number of these which occur in a given time. The slower the rate the lower the pitch: the more rapid the rate the higher the pitch. You are aware that this applies equally to all sonorous bodies, and that, although the pitch of a sound elicited from a vibrating string may be raised to the same extent by either halving its length, quartering its weight, or quadrupling its tension, yet that this is only because by each of these operations the rate of vibration is affected in an equal degree. And when sounds are produced by the vibrations of air contained in tubes, the same effects are obtained by changes in the length and calibre of the tubes, and by the open or closed condition of their ends, as by alterations in the tension, weight, and length of vibrating strings. All that we have to bear in mind then, is, that *high pitch and rapid vibration*, and *low pitch and slow vibration*, are inseparable.

In tranquil normal breathing it is difficult to detect any expiratory sound when listening to the chest-wall; but by giving a very slight voluntary character to the expiratory act, a sound is at once heard. Now this sound will be always found to be of much *lower pitch* than the *in*spiratory sound, if the lungs and air passages are healthy. And the question arises, Why is this? Do not the inspiratory tide and the expiratory tide pass through the very same tubes, through tubes of the same calibre, and therefore ought not each to produce a sound of the same pitch? That they do not produce sounds of the same pitch is due principally to the difference of *rate* in the two currents; although, no doubt, something is due to the altered temperature and volume of the expired air. The inspiratory current is much faster than the expiratory; and as the same volume of air has to be drawn through the tubes in a shorter space of time, the vibrations set up are more rapid, and the pitch is proportionately raised. That such is the case is subject to the simplest proof; for we have only to make a patient snatch a sudden forcible breath, and so increase still further the *rate* of the inspiratory tide, and the pitch will rise proportionately. On the other hand, if a gentle slow inspiration is taken, the pitch will sink; and then by a sharp, forced expiration, the expiratory sound may be made to *rise* in pitch till it is even higher than the previous *in*spiratory sound: or the same experiment can be tried with a common bellows; the pitch of the sound produced by the rush of air through the nozzle can be raised or lowered by increasing or diminishing its rate.

If, then, in the natural state of things, the inspiratory sound is of higher pitch than the expiratory, it is clear that any alteration in this relationship must proceed from some change in the physical conditions, and ought to excite our attention.

It is evident that if the alteration of pitch in either of the sounds is due simply to an increase of rate in the current of air, other things being normal, the duration of the sound should diminish as the pitch increases; for a shorter time must be required to drive the same volume of air through the same tube at a rapid than at a slow rate. We see this exemplified when a chest which is not highly elastic is forcibly expanded, it recoils with suddenness and force: whereas a highly elastic chest expands freely with a less exertion of force, and recoils more slowly and gently. In the first case, the expiratory sound will be short and high pitched; in the second case, longer and of lower pitch. Exactly these two states and their results are seen in the chests of girls and of boys, especially in the front and upper parts of them. I have often watched this when examining large numbers of boys and girls, one after the other, for the Royal Albert Orphan Asylum.

Make a boy take a deep breath while you listen to the front of his chest, and you will find the succeeding *ex*piration short and high-pitched. Make a girl do the same, and the expiration will be longer, softer and lower-pitched; and if, as sometimes happens, you find a boy with a chest like a girl and a girl with one like a boy, the usual phenomena will be reversed.

I may mention, in passing, that this is, in my opinion, the explanation of the much discussed question of the cause of the perceptible expiratory sound in the earliest stages of tubercular deposits in the lungs. The resiliency of the lung is interfered with, the chest expansion is interfered with, and a shorter and more sudden recoil occurs, driving the air through the tubes with more force and at a greater rate than usual, and consequently raising the pitch of the expiratory sound so as to make it audible. A certain rate of vibration is necessary to produce sound at all, and it may be that the rate in healthy normal expiration is not sufficient to be sonorous; whereas, the increase of rate to which I have referred, gives vibrations rapid enough to produce sound. I am not here speaking of that still greater change in the expiratory sound due to obstruction to the outward tide, which occurs in more advanced disease, but of a change in character which long precedes this, and is much more important to identify.

Well then, the first and simplest cause of a rise in pitch in either of the respiratory sounds is an *increase in the rate of current;* and with this, if all else is normal, a *decrease in duration* must correspond.

But suppose the pitch of one of the sounds is raised and at the same time the duration is not decreased, then there must be a diminution in the calibre of the tube through which the current passes. For given two equal volumes of air driven at the same rate through two orifices, either they will pass in the same period of time or the sizes of the orifices must be different. Therefore *an expiratory sound which is both long and high-pitched* must be due to a narrowing of the orifice through which it has to pass.

But at this point a difficulty occurs which for a long time puzzled me very much to explain.

The difficulty is this: How can it happen that a change in the calibre of a respiratory tube, through which both the inspiratory and expiratory currents have to pass, should affect one of these currents more than the other? For instance,—if a bronchial tube is narrowed, why does it not raise the pitch of the inspiratory and expiratory sounds in the *same proportion*, and thus maintain the normal relation between their pitch? The explanation is that which I gave when speaking of backward pressure upon the air-cells, viz., that the *moving powers* in the two cases do not remain the same. In normal respiration, the expiration is performed by the elastic recoil of the lung-tissue, followed up by the elastic recoil of the chest-wall; but directly there is an impediment placed in the way of the expiratory current, the expiratory *muscular* system is called into play, and thus a new element is added to *expiration* which is not added to *inspiration*. An additional force of a new kind drives on the current in one case and not in the other. But with this new element of force comes that other most important one to which I have also referred, viz., that whereas all *inspiratory* efforts tend to expand the chest, and, by taking off the superincumbent pressure, to dilate the air tubes, all *expiratory* efforts tend to compress the lungs, and consequently favor the contraction of the tubes, and this becomes especially the direction in which *muscular expiration* acts; the air-cells which lie upon and in the neighborhood of the air-tubes are pressed upon them. The operation of forced expiration, therefore, acts in the direction to favor and increase the *narrowing* of the expiratory current, and to increase the pitch of its sound, out of proportion to that of the inspiratory sound.

There is a very interesting exception to this rule which serves to prove it. In spasmodic asthma, the *inspiratory* sound is raised in pitch more than the expiratory. This is due to the spasmodic contraction of the muscular fibres of the bronchi, excited by the attempts to inspire fresh air. They, in fact, offer a direct opposition to the normal inspiratory act, and almost paralyze it. They narrow the passage of the inspiratory current, and its sound is raised in pitch, in proportion to the rate at which the inspiratory efforts succeed in drawing it through the narrowed passages. But these spasmodic contractions do not, as a rule, offer the same obstruction to the outward tide; the tubes having been forced to yield to the inspiration, allow the air, now warm and charged with carbonic acid, to escape with less opposition, renewing the vigor of their contraction at every new attempt to draw fresh air through them.

We have then two valuable diagnostic physical signs plainly demonstrated:—

1. A high-pitched long *ex*piratory sound must mean contraction of the respiratory tubes independent of spasm.

2. An abnormally high-pitched *in*spiratory sound, not accompanied by a corresponding change in the expiratory sound, must mean spasmodic contraction of the air passages.*

I will mention another clinical fact which, while it presents a third diagnostic physical sign, confirms the correctness of the other two. If a contraction takes place in a respiratory tube, which is *rigid* in its character, and so situated that it cannot be favored by either inspiration or expiration, the pitch of the inspiratory and expiratory sounds is alike affected by it. This may be best observed in affections of the larger air passages, such as rigid contractions of the openings of the larynx, compression of the trachea and bronchi by tumors, and the like.

Everything connected with the signs and symptoms of disease is beset with sources of fallacy; and these physical signs are no exceptions to the rule.

---

*The great importance of these aids to diagnosis, which I first taught when these Lectures were delivered in 1865, has been since confirmed by daily experience.

But the principal sources of fallacy in this case are not very difficult to eliminate. They consist in the influence of tongues and plugs of secretion adhering to the walls of the tubes, which by temporarily narrowing the passage at the point where the plug is lodged, and by the vibration of the free ends of the adherent tongues, give rise to deceptive elevations in the pitch of one or both of the respiratory sounds. It is evident that these are movable causes; and therefore, before deciding upon the meaning of any alteration in the pitch of the respiratory sounds, the patient should be made to clear the chest by breathing and coughing sharply several times; after which the chest should be examined again. If the alterations of pitch are not essentially changed in position or character by the coughing and sharp breathing, it is pretty certain that they are not due to such movable causes as plugs and tongues of mucus.

In the case of spasmodic contractions, it occasionally happens that the spasm is so capricious that it suddenly gives way during inspiration, and closes upon the expiratory tide; but this reverse of the general rule is but an exceptional case, and is not likely to recur in several consecutive respirations.

Again, we must be on our guard against *transmitted sounds*. A sound or quality of sound generated in one part of the naso-pulmonary tract, may be transmitted through other portions of it. Even changes of sound produced in the posterior nares may be heard down the bronchi; and changes in the quality of sounds generated in the minute air tubes may be heard in the larger passages. The most common of these occurrences, however, is that qualities of sound produced by affections of the larynx are transmitted down the trachea and bronchi, and may be mistaken for changes having their origin there. But really this source of error may be avoided with tolerable ease. A sound must be most intense at the point of generation unless it is reinforced at some other part; and, therefore, when a change is detected in the normal character of a sound, it should be followed with the stethoscope to its points of greatest intensity. It is true that we are again met by a source of fallacy in the different conducting powers of the parts intervening between the source of sound and the chest-surface; but this may be eliminated by resorting to the usual tests of the character of the conducting media. With a little care and a little tact, therefore, we may satisfactorily bring these changes of pitch to which I have referred into the position of *positive diagnostic signs of spasmodic or non-spasmodic narrowing* of the respiratory tubes.

VII. ON POST-NASAL CATARRH.

ALTHOUGH the great majority of Winter Coughs are either cases of Bronchitis, or of Bronchitis and Emphysema conjoined, there is a certain number which do not belong to either of these classes, and require to be carefully distinguished in practice.

I exclude from detailed consideration all cases in which the cough is dependent upon Tubercle in the lungs, as beyond the objects of these Lectures.

It is, however, important to bear in mind that a slight cough coming on with the winter for a second or third time, and nearly disappearing in the summer, may be a sign of the presence of Tubercle in the lungs; and thus "Winter Cough" may be a passing phase in the history of consumption.

It must also be remembered that, as all cases of consumption are accompanied by cough, and as the cough is very apt to be convulsive, they only need the ordinary conjunction of obstructed expiratory tide to become complicated with Emphysema. Emphysema is in fact a frequent accompaniment of Tubercle in the lungs.

Excluding consumption, then, and without attempting to exhaust the subject, the following headings will be found to include the principal exceptional cases of Winter Cough, arranged in the order of their importance:

1. Post-nasal catarrh.
2. Chronic recurrent laryngeal and tracheal catarrh.
3. Ear cough.
4. Follicular disease of the pharynx.
5. Superficial inflammation and serration of the edges of the soft palate.
6. Elongated uvula, becoming relaxed and œdematous with every fresh attack of cold.

These may each exist separately, but they are often found associated in a single case.

The most important of these exceptional cases, in relation to Winter Cough, is Post-nasal Catarrh. It is surprising how often chronic and intractable cough may be found to be due to this simple cause; it is one, therefore, that should never be forgotten when investigating a new case.

The attention of the profession was first specially called to this complaint in a paper which I read before the Abernethian Society of St. Bartholomew's Hospital, in 1854. Subsequent experience of a very large number of cases has fully confirmed the accuracy of the account of the disease which I then drew up, and I therefore give the following extracts from my original paper, with the addition of some matter suggested by more extended observation:—

"Among the diseases which come before the physician, rather as sources of discomfort than as causes of death, the one I am about to describe, and which I have named *Post-nasal Catarrh*, is particularly worthy of attention:—1st, from its frequency; 2d, from the great inconvenience it occasions; 3d, from the serious effects of which it is occasionally the indirect cause; 4th, from its being accompanied by the symptoms of other diseases, and thus misleading the patient, if not the medical man; 5th, from the obstinacy with which it often resists treatment."

"Notwithstanding the obvious existence of Post-nasal Catarrh as a common complaint, it has not, so far as I can learn, ever been described as a distinct affection. I have, therefore, ventured to fill this gap by a brief description of the disease."

Post-nasal Catarrh may be acute or chronic, but it is much the more frequent in the chronic form, and is rather to be classed among the "vestiges" of disease than among primary affections.

The acute form is caused, as a general rule, by recent catarrh superadded to the chronic disease, the latter being the characteristic affection. I have, therefore, selected the chronic form for description. The symptoms are as follows:—

1. A sense of fulness deeply seated in the back of the nose, with a stinging and tingling sensation about the uvula, soft palate, and posterior part of the hard palate. This sensation is much aggravated after sleep, so that the patient *wakes every morning with a sore throat;* but on examination of the throat, no inflammation, ulceration, or swelling is detected.

2. Short tickling cough coming on at intervals, especially night or morning, or if long without food or drink; but on examination of the chest no morbid sounds are present.

3. Frequent hawking and spitting of small pellets of mucus, some of which are not unfrequently of an orange-brown color and very tenacious, and sometimes mixed with clots of blood.

4. On examining the pharynx, shreds of stringy mucus may often be seen hanging down from behind the velum; or the back of the pharynx is coated with grey, or brownish, or green adhesive mucus, and sometimes, but not always, the mucus follicles are enlarged and red. In some old-standing cases the green mucus dries like a scab on the back of the pharynx, and in the worst cases the breath is insufferably offensive.

5. The morning sore throat is not relieved until, after much forcible blowing of the nose, which is usually dry, a few pellets of inspissated mucus are

removed. This mucus is often orange-brown from the admixture of a very small quantity of blood. The excretion of these pellets, which is a very troublesome process, gives immediate relief; very often distressing retching is produced by the presence of the mucus between the posterior nares and the pharynx, too low to be affected by nose blowing, and in these cases warm drinks and swallowing mouthfuls of bread assist in giving relief. But a certain amount of stinging and tingling soon returns, and is especially teasing when any mucus collects on the back of the uvula. This continues through the day, to be aggravated as before by the succeeding night's rest.

6. In order to relieve the uneasiness about the fauces and posterior nares, the patient is constantly annoyed by an almost irresistible desire, either to draw the mucus down the throat by a forcible inspiration, or to force it into the nose by an opposite effort; and, therefore, those who suffer from Post-nasal Catarrh acquire a habit of making a peculiar noise in the nose and throat, which is *pathognomonic of the complaint in its chronic stage.* It is produced by inspiring by the mouth, closing the glottis, and then, with the tongue pressed against the hard palate, suddenly opening the glottis and jerking a gust of air up the pharynx through the posterior nares into the nose. This gives temporary relief. Considerable comfort is also obtained by swallowing food.

7. The usual symptoms of Coryza—watery discharge from the nose, nasal voice, feverishness—are absent. If by any cause these symptoms are produced, those of Post-nasal Catarrh are for the time somewhat relieved, but only to be aggravated afterwards.

The history of the case will generally show that the post-nasal affection has been left behind as a vestige of one or more severe attacks of Influenza, Coryza, or Quinsey, or of many slighter Catarrhs coming in quick succession. In many cases I have found the disease traceable to a long-past attack of one of the acute specific diseases, especially small-pox, measles, or scarlet fever; and children who have suffered from Syphilitic snuffles in early infancy are very apt to be the subjects of Post-nasal Catarrh as they grow up.

Although the mucous membrane of the posterior nares and fauces is usually implicated, the special seat of this affection appears to be in the sphenoidal and posterior ethmoidal cells. The following considerations lead to this opinion:—

1. The deep seat of the sensation of fulness.

2. The absence of nasal voice, and of interference with nasal respiration, except when the secretion has been voluntarily forced into the nose, or when ordinary Catarrh is superadded.

3. The great difficulty in dislodging the secretion and of bringing it within the range of a sneeze.

4. The slight interference with the sense of smell, and the tendency of the secretion to accumulate and inspissate during the time the patient is in the recumbent posture.

5. The tendency of the secretion to flow down by the posterior nares, rather than by the anterior; which corresponds with the direction of the superior meatus, into which the sphenoidal and posterior ethmoidal cells discharge themselves.

6. The stinging and tickling of the uvula, and hard and soft palates, unaccompanied by any constant morbid appearances in them; which corresponds in position with the distribution of the spheno-palatine branches of the superior maxillary nerve, and may thus be accounted for as the transferred impression of irritation in the sphenoidal cells.

7. It seldom develops all its characteristic symptoms before the period of puberty, which corresponds with the development of the sphenoidal cells.

The relief to the cough by taking food or drink is simply due to the removal, by these means, of the mucus clinging to the back of the fauces.

The duration of the complaint, when once established, appears to be quite indefinite, unless removed by treatment, and its tendency is to get gradually worse.

Among the serious effects of which this affection is sometimes the indirect cause, may be mentioned the production of hernia, the rupture of blood-vessels, strains to the lungs, Emphysema, and injuries to the internal ear, from violent nose-blowing.

True Post-nasal Catarrh must be distinguished from the simple post-nasal discharges, so common with children, which are of quite a different nature. They appear to be simply due to the difficulty which children have in effectually blowing their noses; hence the nasal secretions, when augmented by cold, escape by the posterior as well as by the anterior nares, and during sleep almost entirely by the former. This discharge into the fauces, as in true Post-nasal Catarrh, gives rise to troublesome cough which, through neglect of auscultation, I have often seen mistaken for Bronchitis, and much time and treatment wasted under this impression.*

---

* A considerable number of cases of supposed "consumption" are brought to me every year, which, on careful examination, turn out to be nothing more than chronic post-nasal catarrh, with sanguinolent sputa.

VIII. ON EAR COUGH.

MY FRIEND, Dr. Cornelius B. Fox, has kindly furnished me with an abstract of a valuable paper, read in the Physiological Section at the Annual Meeting of the British Medical Association at Leeds, July, 1869, in which he first called the attention of the Profession to this variety of cough. From my own experience I can confirm his observations.

For example:—A distinguished architect consulted me, December, 1870, in so serious a state of exhaustion and emaciation that he was on the point of relinquishing his profession. He had a frequent irritating cough, and could not take food without agonizing pain. The cough and emaciation might easily have led to the conclusion that he was in a consumption, and such was the opinion of his friends and of himself. On examining the chest I found no disease of the lungs whatever, and close investigation of his symptoms showed that the wasting was due to want of food, the pain after taking food being so severe that he abstained from eating rather than incur the suffering. Pancreatic emulsion, two teaspoonfuls one hour after breakfast and dinner, acted like a charm upon his apepsia, enabling him to feed without pain almost at once; and he came to me shortly after his first visit saying that he had no pain and would be quite well but for his cough, which did not get a bit better. Upon this I again made a careful examination, and found his chest, larynx and throat quite healthy; but I discovered that one ear was a little deaf, and had a slight chronic discharge from it; and on introducing the ear speculum to examine the drum, a fit of coughing was at once excited, and the patient cried, "Oh! you must not do that; I never can touch that ear without setting up the cough." And such was the truth. I found that every time the ear was touched the cough came on. The external meatus was inflamed and irritable quite down to the margin of the tympanum. I ordered blisters behind the ear, sedative lotions into the meatus, and the orifice to be kept softly blocked up with cotton wool. The cough entirely disappeared, and with the continuance of the pancreatic emulsion after food, for some weeks, the patient completely regained health and flesh, and two years afterwards was actively pursuing his profession and enjoying his life.

Abstract of Dr. Fox's Paper on Ear-Cough.—Cough is most commonly symptomatic of diseases of the lungs and their coverings, of the trachea, bronchi, larynx and fauces. Books tell us that it is occasioned by affections of the tonsils, uvula, pharynx and neck. Organic diseases of the thoracic viscera, diseases of the œsophagus, spine and spinal cord, affections of the heart, liver and stomach, all have occasionally a cough as one of their symptoms. The irritation produced by teething, by enlargement of the bronchial glands, by worms in the intestines, by tumors, aneurisms, tubercles in the lungs, will produce cough. The inhalation of the dust of ipecacuanha, and of certain animal and vegetable emanations, will sometimes induce cough. Then there are whooping cough, asthma, nervous or hysterical cough, and the cough occasioned by the presence of foreign substances, either solid, liquid, or gaseous, in the air-passages. Uterine derangements, irregular gout, and an accumulation of bile in the hepatic ducts or gall-bladder, are all included amongst the causes of cough.

Lastly, there is a cough caused by irritation of the auditory canal—and that only in some people—to which I have given the name of *ear-cough*. I should not have presumed to name it, were I not pretty sure that this kind of cough has hitherto escaped description, and even recognition in our text-books. And this fact is the more singular, inasmuch as the sympathy between the auditory canal and the larynx was well known to the older writers, although apparently lost sight of by modern authors. This kind of cough has doubtless been confounded, up to a very recent period, with nervous cough, which occurs in persons of highly nervous temperament, and is due to a convulsive action of the throat muscles; or else it has been included in that *terra incognita* of idiopathic coughs.

One of the problems of Cassius Medicus was the following:—"Why does irritating the ears, as, for example, with a speculum, cause sometimes a cough, just as if the trachea were irritated?" Whytt, in his work on the *Sympathy of the Nerves*, published in 1767, refers to it, and states that, when the trachea has been rendered more sensitive than usual by a catarrh, cough is more readily produced by irritation of the auditory canal. Pechlin (*Observationes Medicæ*, Lib. ii., No. 45) affirms that "an irritation of the meatus auditorius will often excite coughing and sometimes vomiting." Coming down to more recent times, we find that Kramer, in his treatise on the *Diseases of the Ear*, published in 1837, makes the following solitary observation relative to this subject: "Tickling and scratching the meatus excite in the larynx a troublesome inclination to cough." Romberg states that "Pruritus of the external meatus auditorius, from hyperæthesia of the auricular branch of the vagus, is sometimes observed and is accompanied with cough and vomiting." The only references to this sympathy, with which I am acquainted, in recent works on medicine, are the following: Dr. C. J. B. Williams, in his *Principles of Medicine*, whilst enumerating the reflected and sympathetic sensations, writes: "Touching the external auditory meatus causes a tickling sensation of the epiglottis." Toynbee, in his work on *Diseases of the Ear*, says: "In certain cases the presence of a foreign body in the meatus gives rise to coughing, and even to vomiting; symptoms which seem traceable to irritation of the auricular branch of the vagus nerve." Yearsley refers in general terms to the alterations of the voice, as regards its pitch and quality, which occur in cases of deafness dependent on diseases of the ear, but does not otherwise allude to this connection.

In my graduation thesis, "Concerning the Laryngoscope and some Laryngeal Diseases" (June, 1864), presented to the University of Edinburgh, reference was made to the sympathy subsisting between the external ear and the larynx, and an explanation of the same was advanced. In a paper written by me in 1868, entitled "The Sympathy between the Auditory Canal and the Larynx," the occasional occurrence of ear-cough was adverted to. The advisability of a careful examination of the auditory canal was also urged in all cases where no affection of the respiratory tract can be discovered, and in which an obstinate cough, whether laryngeal in character or otherwise, obtrudes itself as a prominent symptom. Further observation in this country and on the Continent has convinced me that a state of hyperæthesia of the nerve supplying the external auditory meatus is not of unfrequent occurrence, and that a cough, solely dependent on the existence of some irritation in that canal, is by no means rare.

With the object of ascertaining the percentage of those subject to this sympathetic peculiarity, 108 persons have been examined by myself, and others under my direction, with the following results:—Males examined, 37; females examined, 45; sex not noticed, 26; total, 108. Cases in which a sensation of tickling in the throat, and cough, were occasioned by a titillation of the auditory canal, 22; cases in which nausea was *alone* produced, 3. In one of the cases of irritation of the throat with cough nausea was also complained of; and in another of that number vomiting was said to be sometimes produced. It is my impression that both ears display this extreme susceptibility to impressions more frequently than one ear only. In twelve cases, the ear was noted by the titillation of which these symptoms could be excited. Of these, in seven cases the left ear alone, and in the remaining five both ears, exhibited this peculiarity. I have hitherto only seen one case where the right ear was solely affected with this hyperæsthetic condition.

Dr. Denton, who has kindly assisted me by the examination of many dispensary patients, was somewhat astonished to find that, in one woman, about twenty-four years of age, whom he examined, vertigo was alone complained of.* In another woman, about thirty years of age, a sensation of tickling in the throat, cough and nausea, were produced by him on irritating the left ear; whilst vertigo was only experienced when the right ear was experimented on. He informs me that in neither of these cases was there any hysteria or fanciful nervous condition. He believes, moreover, that the statements of these women are thoroughly reliable.

The curious hyperæsthetic condition which we have been considering would seem, then, to be present in about twenty per cent. of those examined. I have not yet sufficient data on which to form an opinion as to its relative frequency in the sexes.

The response to the inquiry, as to the duration of the peculiarity, was generally to the effect that they had been aware of a feeling of irritation in the throat, usually followed by a cough on employing the ear-pick, so long as they could remember.

It is possible that this state of hyperæthesia may be

---

* Dr. Brown-Séquard, in his *Physiology of the Nervous System*, refers briefly to the production of vertigo by an irritation of the neighboring auditory nerve.

present more frequently than the foregoing percentage indicates, because those submitted to examination were nearly all workhouse and dispensary patients. In these people, the nervous organization is not, of course, so sensitive as in the higher and better-bred classes of society, and their powers of observation are more limited. I speak of those whose attention is rarely directed to any sensation short of pain, or of that which prevents the employment of the limbs and the ability to work.

Here are some short notes of a case of this hyperæsthetic condition of the nerve supplying the auditory canal, which differs from the majority of these cases, in possessing the peculiarity on one side only.

Case I.—A. B., a professional man, of middle age and nervous temperament, healthy, but somewhat overworked mentally, has been annoyed, so long as he can remember, with a feeling of irritation in the larynx whenever he has introduced an ear-pick for a short distance within his *left* auditory canal. This sensation frequently excites a violent cough, of a suffocating or convulsive character. The organs of hearing appear, on examination, to be perfectly healthy, the sense being somewhat more acute on the affected side than on the other. A careful examination of his larynx by means of the laryngoscope, has assured me of the absence of any abnormality whatever. He often finds that the excessive use of his voice, as in long-continued singing, produces a pain in the ear, which extends into the zygomatic fossa and along the lower jaw towards the chin.

Irritation of the auditory canal in those who are the subjects of this hyperæsthetic peculiarity, does not *always* produce a cough. The situation of the irritation and its intensity have much to do in the production of this symptom. I know a medical man who is now troubled with a chronic inflammation of the dermis lining the meatus, accompanied by desquamation of the epidermis. This affection being limited to the *outer part* of the auditory canal, being mild in character, and free from symptoms of any severity, produces simply a sensation of tickling in the throat.

I will now give brief abstracts of two or three of the cases of ear-cough which have come under my notice.

Case II.—M. H., a healthy-looking married woman, aged 50, of sanguineous temperament, applied on account of a cough, in the endeavor to remove which she had spent much and profited nothing. She had suffered from it for eighteen months, during which period it had varied in force and frequency. At the time of her application, the cough seemed to be becoming worse. Her voice was unaltered. The cough was laryngeal in character and most distressing. Failing to discover any cause for it on carefully examining the lungs and other viscera, I illuminated her larynx by means of a Tobold's condenser, and obtained a good view of the vocal organ by the aid of the laryngeal mirror. Nothing abnormal could be detected, if a slight exaggeration of the usual rose tint of the parts be excepted. Observing that she was somewhat deaf, I made some inquiries respecting her ears. She informed me that her right ear had given her some annoyance for nearly two years. The symptoms of which she complained were those of an accumulation of cerumen. She had, from childhood, noticed that, on cleansing her ears, a feeling of irritation in the throat and a cough were often excited. On making an examination of the auditory canal, a large plug of hardened wax was found, and removed by means of injections. The appearance of a small quantity of purulent discharge led to the discovery of a small oval ulcer on the floor of the canal, about one-eighth of an inch from the tympanic membrane. After two or three applications of a solution of nitrate of silver to the spot of ulceration, cicatrization was complete, and in a few days she was not only entirely free of the aural affection, but of the troublesome cough it had occasioned.

Case III.—N. W., a young lady of nervous temperament, aged about 22, of somewhat anæmic aspect, consulted me respecting her ears, as she suffered from deafness. Without entering into the history of the affection, which came on as a sequel to scarlet fever, it will be sufficient for me to state that the Eustachian tubes were the seat of chronic inflammatory action leading to obstruction. She had been deaf for two or three years, and had apparently exhausted the resources of orthodox medicine in the part of the country where she resided. Under these circumstances, her parents, with some considerable hesitation, allowed her to consult an advertising quack. This man ordered her to employ some drops, which appeared to be composed of spirit of cajeput oil, or some other spirituous irritant. The result of this application to the auditory canal was the production of a great amount of irritation, closely resembling eczema. This irritation, which seemed most intense along the floor of the canal, was accompanied by a cough. I need hardly say that, as soon as the young lady came under my care, these pernicious drops were thrown away, and my sole endeavor was to repair the mischief occasioned by them. The cough could not be accounted for by any visceral affection, and was ascribed by me to the irritation of the nerve supplying the auditory canal, in one who was the subject of the peculiarity already adverted to. In about three weeks, the eczematous condition of the external auditory meatus had disappeared, and the accuracy of my diagnosis was established by the subsidence, *part passu* of the cough.

Case IV.—Toynbee, in his *Diseases of the Ear*, refers to a patient under his care, who suffered from a most intractable cough. He had a portion of dead bone in his auditory canal, which was removed. The withdrawal of this source of irritation was attended with an immediate disappearance of the cough, which no remedies had been able to subdue.

Before entering on the consideration of the mode in which ear-cough is produced, it will be necessary to make a few remarks with reference to the nervous supply of the auditory canal. It would seem that there have been erroneous ideas prevalent as to the source whence its nerves are derived. Romberg, whose opinion is endorsed by the late Mr. Toynbee, states that the auricular branch of the vagus nerve is distributed to the external auditory meatus, and that this nerve is concerned in the production of a cough when that tube is irritated.

Now, the best anatomists inform us, and their views have been confirmed by my own dissections:

1. That the auditory canal is supplied with nerves from the auriculo-temporal branch of the inferior maxillary division of the fifth cranial nerve. They are two in number, and enter the interior of the meatus between the osseous and cartilaginous parts.

2. That the auricular branch of the vagus is one of the several nerves which find their way to the external

ear, this particular nerve being distributed to the posterior part of the pinna.

The other cranial nerve which takes part in the production of the sympathetic phenomenon that we are considering, is of course the vagus, which alone supplies the larynx by means of its superior and inferior laryngeal branches. It is well known that an impression at the peripheral extremity of a sensitive nerve may produce such a change in that part of the nervous centre from which it arises, as to excite a motor or sensitive nerve implanted near to it; and that, if a sensitive nerve be stimulated at its origin, a sensation is produced which is referred to its peripheral extremity. As examples of reflected sensations, may be instanced: *a*. The otalgia which often accompanies a toothache, and which is relieved by the introduction into the auditory canal of a little laudanum or chloroform; *b*. The pain over the brow sometimes induced by the swallowing of ice or cold water, or by a derangement of stomach digestion.

If the sensation which is reflected be powerful, a reflex action is sometimes excited in consequence of the irritation induced. The feeling of irritation in the larynx, as a result of the titillation of the nerve distributed to the auditory canal, which feeling of tickling often provokes a cough, supplies us with an example. The impression produced in the ear in those amongst whom this sympathy between the ear and the larynx is exhibited, is probably conveyed by the auriculo-temporal branch of the inferior maxillary division of the fifth cranial nerve to the deep origin of its sensitive root, which is in close proximity to the deep origin of the vagus in the floor of the fourth ventricle. Here, a change is in all probability effected, which results in the stimulation of certain of the sensitive fibres of the vagus nerve. A sensation is produced by this stimulation, which is referred to the peripheral extremity of the superior laryngeal or sensitive nerve of the larynx. If an oft-repeated or powerful sensation is reflected, the irritation induced excites the reflex action of coughing, to free the larynx of the *supposed* irritation.

The cough arising from an irritation of the dental branches of the fifth nerve, which may occur at any age, but is generally seen in children who are teething, is occasioned in a precisely similar manner.

In conclusion, my observations may be thus summed up:—

1. From amongst the unknown group of idiopathic coughs, may happily be rescued from obscurity a cough which is excited by an irritation of the meatus auditorius externus in certain individuals.

2. The persons referred to are those who possess a hyperæsthetic condition of the nerve supplying that

canal, and in whom any slight titillation of this nerve induces a feeling of tickling in the throat.

3. This hyperæsthetic state generally exists in both ears, sometimes, however, only in one, and occurs in about twenty per cent. of those examined.

4. Its existence can usually be traced back to childhood, and is probably a congenital peculiarity.

5. The nerve of the ear concerned in the production of ear-cough is not a branch of the vagus, as Romberg and Toynbee have affirmed, but is a branch of the auriculo-temporal branch of the fifth cranial nerve.

6. This sympathy between the ear and the larynx is an example of a reflected sensation, in which the connection between the nerves involved takes place in the floor of the fourth ventricle.

7. Vomiting is occasionally, but rarely, the result of the application of an irritant to the nerve distributed to the auditory canal.

IX. THE NATURAL COURSE OF NEGLECTED COUGH.

A VERY important practical question will be frequently asked you by patients suffering from Winter Cough, a question which it will be your duty to answer with perfect candor, and with such caution and consideration that you may not be misunderstood; for issues of the greatest concern to your patients may hang upon your answer. The question is this—"What will be the consequence of neglecting my cough?"

I need not tell you that the answer to this very natural question will differ materially in different cases, and must depend upon a carefully balanced consideration of many points, of which the following are the principal:—

1. The age, sex, and occupation of the patient.
2. The hereditary or acquired diathesis.
3. The clinical group to which the case belongs, viz., whether (*a*) it is a case of Emphysema, with a history of previous Bronchitis, but at the time of examination free from Bronchitis; or (*b*) a case of Bronchitis with no Emphysematous complication; or (*c*) a

case in which both Emphysema and Bronchitis exist; or (*d*) one of those exceptional cases in which there is neither Bronchitis nor Emphysema.

4. Whether or not there is a great susceptibility to Catarrh, and which portion of the naso-pulmonary mucous membrane is most readily attacked, *i. e.*, whether the colds affect first the nose, the throat, or the chest; and whether a cold caught in one portion of the tract is accustomed to expend itself there, or to travel rapidly to some other portion, and which part is most subject to such secondary attacks.

5. What are the means and opportunities of the patient with regard to the extent and completeness with which necessary treatment can be carried out.

The fact is that many persons who have a Winter Cough have got sufficiently used to its discomforts, and suffer so small an amount of chest pain, that if they can be assured that no harm will come of allowing it to go on from year to year, they would rather leave it alone than undergo the trouble, restrictions, and expense, necessary to its cure; and I think we are bound, as medical men, to carefully consider the convenience and interests of patients in these respects before allowing them to commence a course of treatment.

This will be especially necessary with patients of limited means, dependent upon their persistently following their occupations; that large class of persons who can neither afford to keep at home even in the worst seasons of the year, nor to indulge in changes of climate; and from the nature of their pursuits and disabilities, they naturally constitute the largest class of sufferers from all the effects of Naso-pulmonary Catarrh.

Let us then enumerate the ordinary course of Naso-pulmonary Catarrh when allowed to take its chance without treatment either by medicine, regimen, or climate, in a person not specially influenced by hereditary predisposition.

1. For one or more years a great tendency to colds in the head after exposure to wet, cold or wind, and frequently without such exposure, from mere barometric changes in the atmosphere. The colds gradually becoming more tedious and difficult to remove, till at length one cold has hardly disappeared before it is reinforced by another.

2. A gradual extension of the colds in the head down the throat, or a sudden tendency to attack the throat first instead of the nose.

3. A gradual extension of the colds in the head or throat down the larynx and trachea into the bronchial tubes; or a sudden tendency to attack the air tubes first instead of the throat or nose.

4. Supposing none of the attacks to have been very

severe or protracted, these catarrhs in the nose, throat or chest clear off completely, so that when free from cold for some time, under fortunate circumstances, and in fine, warm weather, the patient appears quite well, no short breath or cough remaining.

It must also be observed that all the effects of the gradual progress described in headings 1, 2, 3, 4, may be produced at a blow by a single very severe and protracted catarrh.

5. From the frequent repetition of the above states, or through some unfortunate circumstances having induced one or more attacks of unusual severity, some of the bronchial tubes become dilated, the naso-pulmonary membrane becomes so much thickened that it does not recover its normal state in the intervals of attack, or even under the influence of summer weather; and thus the breath remains permanently short, and there is an habitual cough and expectoration. Both the short breath and cough in this stage are much improved during summer weather, but do not actually depart.

6. It is from this point that the more serious consequences of Winter Cough begin to date. The dilated tubes keep up an offensive and wasting expectoration. The thickened membrane, and consequently diminished calibre of the air tubes, permanently obstruct the freedom of respiration; less complete and rapid

changes occur in the residual air of the chest; the vascular circulation upon which the aëration of blood depends is impeded, and an abnormal tax is put upon the right heart in its duty of propelling the blood through the lungs. Thus retrograde venous congestion of all the organs in arrear of the right heart gradually increases, and the right heart itself hypertrophies to meet its increased work.

7. The obstructed expiratory tide keeps up a constant undue pressure backwards upon the air-cells, the elastic walls of which gradually yield to this, and the capacity of the cells becomes abnormally enlarged. This is greatly aggravated by the frequently recurring attacks of cough, consequent upon the other conditions, and the cells become stretched beyond their power of recoil; that is to say, they become permanently Emphysematous.

8. The attenuation and over-stretching of the walls of the air-cells interfere with the free circulation of blood in their capillary vessels, and add another to the already existing interferences with blood aëration, and another to the already existing causes of hypertrophy of the right heart, and of retrograde venous congestion.

9. A condition now exists in which every day produces new mischief of the most serious character. The over-stretched, ill-nourished walls of the air-cells

degenerate, and lose thereby still more of their power to resist the distending influence of the backward pressure of air; and becoming, as it were, rotten in their texture, they tear and break up under the shock of the convulsive coughs common to this stage of the complaint. The permanently obstructed venous circulation, after deranging the action of the stomach, liver and kidneys, leads to diseases of these organs, culminating in dropsy.

Such is a fair sketch of the gradual progress of a neglected Winter Cough, due to Naso-pulmonary Catarrh. In its course many accidents may occur to which I have not alluded. As a matter of fact, a case seldom runs on through all its stages without such accidents, and when they occur they precipitate all the evil consequences, or probably terminate the patient's life before the case has run through all its usual stages.

One of these accidental complications is so full of peril, and, unfortunately, so apt to occur, that I must stop to caution you against neglecting it. Every now and then an attack of Bronchial Catarrh will be accompanied by active congestion of portions of lung-tissue, and if the patient is predisposed to tuberculous disease, and happens at the time to be in depressed general health, there will be great danger of the occurrence of tuberculization in the congested tissue;

and I need not tell you that we are then brought face to face with all the dangers and difficulties of Consumption, in addition to those of the Winter Cough. But independent of a tuberculous diathesis, we must always regard these attacks of active congestion of portions of lung with great apprehension, and lose no time in dispersing them; for their repetition in the same part, or their accidental occurrence with unaccustomed severity, will lead to disintegration of the affected tissue, and although unaccompanied by any deposit of miliary tubercle, the patient's life will be placed in great jeopardy by a chronic wasting disease, having many of the characters of tubercular consumption.

While, therefore, you should avoid any approach to unnecessarily alarming your patients, it will be your duty, as I said at first, to candidly inform them of the dangers with which the natural course of their complaint is beset when left without treatment.

We must not, however, forget the other side of the story. Notwithstanding the unpromising picture I have just drawn, cases occur, every now and then, in which I am accustomed to advise my patients to let their Winter Cough alone. For example:—

A patient tells you he has had a cough nearly every winter for five, ten, or twenty years—that it does not get worse from year to year, but, on the contrary, has never been anything like as bad as the first year of its occurrence—that the breath in summer is not observably short, and that the cough always disappears with the warm weather. His health has not apparently suffered, and he has never staid at home on account of his cough. On examination, you find a small amount of chronic thickening of the naso-pulmonary mucous membrane, a small amount of bronchial secretion of a loose character, easily expectorated; no heart disease, no Emphysema; the cough does not seriously interrupt his sleep, and does not occur in violent convulsive fits; there is no great sensitiveness to colds, and no marked diathetic condition. You learn also that he is quite unable to leave his daily occupation and get change of climate, or even to stay at home in bad weather; that he has frequently had advice and medicines for his cough, but without material benefit, and expense is a serious consideration with him. Finally, you find that the cough dates from one severe neglected and protracted attack of naso-pulmonary catarrh, occurring in the midst of good health, and traceable to a great and unusual exposure—that, in fact, it was not due to any diathetic predisposition, but to an accident in the conditions of life, and that the damage to the naso-pulmonary mucous membrane is rather the vestige of that one attack than a gradually encroaching condition due to repeated attacks.

Under these circumstances, if you find, on careful enquiry, that the advice and treatment he had previously received were good, and such as ought to have benefited him if his complaint were easily amenable to the effects of medicine alone, you will do well to give the patient instructions for dress and other precautions against fresh colds, advise him to keep a careful watch as to any signs of either his cough or his breathing getting worse, and in that case to apply at once for treatment; and to take advantage of the first chance of a winter in a more suitable climate; but, pending these contingencies, not to spend either time or money on the treatment of his Winter Cough.

PART II.

THE TREATMENT OF COLD, COUGHS AND CONSUMPTION.

I. THE PATHOLOGICAL CONDITIONS IN WINTER COUGH.

WE now come to the all-important question of treatment. What are we to do when a case of Winter Cough comes before us in practice? Let us consider what it is that we have to treat. First and most prominently thrust on our attention is the Winter Cough. That is clear enough; and it must, I think, be equally clear to us all that, if we direct our attacks upon the Cough, as though that were the disease, we shall make a fatal mistake, and shall most certainly be disappointed in our hopes of doing any permanent good.

What we have to treat, then, in the large majority of cases of Winter Cough, is a combination of some or all of the conditions which I have already discussed, viz:—

1. Dilated right heart.

2. Collapsed lung.

3. Emphysema.

4. Thickened naso-pulmonary mucous membrane, with narrowing of the air-passages.

5. Catarrh of the naso-pulmonary mucous membrane, of greater or less extent.

6. An undue susceptibility of the naso-pulmonary mucous tract.

7. Local and general conditions, favoring or producing susceptibility of the mucous membrane.

8. Cough and short breath; symptoms of the existence of the conditions already enumerated.

9. Dilated tubes.*

10. Disintegration of lung tissue.†

1. Dilated right heart. This is a secondary affection, produced by the persistence or frequent repetition of obstructed pulmonary circulation. It is a complication so sure to occur in a protracted and neglected case, and, once established, it exercises so important an influence over its future course, that we ought never to forget the tendency to its occurrence. During the whole course of any case of Winter Cough which comes under our care, we must remember con-

---

*In dilated tubes, the inhalation of vaporized carbolic acid is of especial service.

†In the treatment of chronic localized catarrhal disintegration, I have found a *seton* over the affected part a very potent remedy.

gestion of the right side of the heart as a thing to be continually guarded against. If no organic change has yet occurred in the organ, we have to remember this as a thing to be prevented; if the organ is already hypertrophied or dilated, or the tricuspid valve has already ceased to prevent regurgitation, we have to remember that all these conditions will be aggravated or kept in check in proportion as we guard against congestion of the heart. We have to remember that congestion of the right heart is the common cause of Dropsy, and that this miserable complication of a Winter Cough will come and go as we permit or prevent stasis of the pulmonary circulation. And I need not remind you of the damaged liver and the damaged kidneys which, in time, result from the neglect of retrograde venous congestion.

It would be impossible in this lecture to enter into the details of the treatment of dilated heart. Let me simply warn you, in passing, that in every case of Winter Cough danger signals must be placed upon the pulmonary circulation, the right heart, and the great veins.

2. Next on our list stands Collapsed Lung. In the course of a case of Bronchitis, Whooping Cough, or any other affection of the air passages, whenever a portion of lung is suddenly deprived of the power to be inflated, by the presence of a plug of secretion

sucked into its main air passage, a set of symptoms is produced, the severity and importance of which will vary with the obstinacy with which the obstructing plug resists the attempts at its removal, and with the extent of lung-substance cut off from the pneumatic circulation.

Portions of lung are frequently being temporarily blocked off in this way during the fits of coughing which attend naso-pulmonary affections; but the plugs are so quickly removed and the admission of air restored, that, in the majority of instances, no damage remains. But every now and then it happens that the plug fails to be removed. It is just sufficiently dislodged by expiratory efforts to allow the air-cells behind it to be emptied, but is borne back into its place on the front of the inspiratory tide, and fixed in the threshold of a portion of collapsed and useless lung-tissue.

Under any circumstances, however small the portion of lung thus put in peril, whether the plug is eventually removed or not, a very distressing convulsive cough is set up, which, as I have already stated, may cause Emphysematous distension of the air-cells; and when an important branch of the bronchi is in question, and a large portion of lung-tissue at stake, unmistakable and most alarming symptoms attend the accident. It is not my purpose, however, to treat here of these symptoms further than to remind you that, in this way, the occurrence of collapse of portions of lung may happen to constitute an important feature in a case of Winter Cough. But what we have to consider here is the existence of a portion of lung, thus damaged and useless, in its position as one of the possible accompaniments of a Winter Cough, and to what extent, if at all, it need influence our treatment.

Now, in this capacity it may soon be disposed of. First, you have to take care when making your physical examination that you do not mistake it for a portion of hepatized lung, and thus waste your own and your patient's time in treating it as such; and, in the next place, having made up your mind as to its nature and the probable length of time it has existed, there are two courses to be followed:—First, if it is at all recent, cautiously to endeavor by rational measures, which will be chiefly gymnastic, to restore it to a permeable condition; and, second, if it is clearly of old standing, to let it alone. So great is the power of the organism to compensate such damages, that considerable portions of lung can be deprived of function without producing more than temporary distress.

But what I wish particularly to impress with regard to the discovery of a portion of collapsed lung is this, —to bear clearly in mind the mode of its production to remember that what has happened once may happen again to the same patient,—that every time these violent convulsive coughs are set up you run the chance of having a fresh set of air-cells deprived of their elasticity by the over-distension of their walls during the expiratory shock,—and that the cause of all this is the naso-pulmonary Catarrh by which the plug was produced that stopped up the tube. Of this Catarrh I shall speak hereafter.

3. Emphysema. I have shown you that Emphysema may come either before or after the first history of Winter Cough, although in the large majority of instances it comes after the cough. I have shown that Emphysema, *per se*, cannot be considered as the cause of the cough, because the Emphysema may exist without the cough, and the cough may exist just as much without as with the Emphysema. The only way in which the Emphysema can be considered as a cause of cough is by its predisposing to affections by which the cough is produced. Practically it rarely happens that Emphysema exists for any length of time without some circumstances arising which bring on a cough.

I have shown you that when the commencement of Emphysema has preceded the commencement of Winter Cough, it may be—nearly always—attributed to one of the following causes:—

1. Violent expulsive acts.

2. Violent exertions of force in lifting or carrying weights.

3. Convulsive fits of Cough; as in Croup, Whooping-cough, Laryngitis, and the like; in which the cause of cough is a temporary one, the cough ceasing but the Emphysema remaining.

4. Violent fits of sneezing and of nose-blowing, under peculiar conditions of obstruction.

It is evident that, in the fourth of these sets of causes, the conditions likely to cause the sneezing and the nose-blowing will usually be such that, if they continue for any length of time, they will become causes of cough; and thus the occurrence of Emphysema before the cough, instead of after it, is accidental.

With the first three sets of causes, it is evident that they may have ceased for any length of time after producing the Emphysema; and, except so far as the existence of Emphysema is concerned, they may have no possible influence over the after history of the case. A man may, for example, have overstrained his lungs by lifting too great a weight twenty years ago; or a child may have had Whooping-cough and overstretched the air-cells during one of the paroxysms; and the man and the child may have remained Emphysematous ever since, although the Whooping-cough had long been completely cured, and the heavy weight had never again been lifted.

In these cases, the Emphysema may be considered as a disease *per se* as long as it remains uninfluenced by superadded diseases. It is well, then, for our present purpose of correctly estimating what it is we have to treat, that we first consider what are the essentials of this simple Emphysema.

In the first place, it consists of a portion of lung the air-cells of which have been *over-stretched;* and it is important to recollect what this over-stretching means when it exists in its smallest appreciable degree.

You know that in the healthy state the walls of the air-cells are elastic, and that it is the essential condition of perfect elasticity that a body shall, when stretched, have the power of recoiling to exactly the same position in which it was before the tension was applied to it. In proportion as it loses this capability, it deviates from the standard of perfect elasticity. A perfectly elastic body will retain this property until it is so far stretched that it snaps; but most elastic bodies retain their elasticity within narrower limits, and lose the power of perfect recoil before they sufficiently lose cohesion to snap; and thus they may be stretched to some extent beyond the point at which they retain the power of perfect recoil. It is this which constitutes over-stretching of an elastic body; and the elastic air-cells of the lungs are susceptible of a certain amount of this over-stretching without rupture.

The elasticity of the lungs is so nicely calculated to meet the requirements of respiration, that, in the normal condition, after the fullest normal inspiration, the lungs recoil by elastic force alone to a condition of fullest normal expiration, and yet retain an elastic power competent to recoil still farther; so that when full ordinary or elastic expiration is supplemented by extraordinary or *muscular* expiration, the lung is still recoiling before the contracting chest-wall. Thus, in the natural state, *pressure of the chest-wall upon the superficies of the lung is unknown.*

Upon the perfection of this degree of elastic recoil everything depends.

The smallest appreciable degree of *over*-stretching of the air-cells may be taken to be such as shall leave their power of recoil intact to the extent of full *ordinary* expiration, but deprive them of that which should still carry them on before the contracting chest-wall in extraordinary or *muscular* expiration.

But small and trifling as this amount of over-stretching and of impaired function may at first sight appear, it constitutes the first step in a most important series of changes, and for that reason must take rank as a very serious disease.

It is true that, when the loss of perfect elasticity is thus limited, there is nothing in it which need at all interfere with *ordinary* respiration. The lung can expand when the chest-walls are expanded by inspiration, and recoil before them when they recoil. But all the normal calculations are deranged when any cause arises which requires *muscular* expiration to be put into force. Then, at once, a totally abnormal condition is discovered, in which the chest-wall has to exert pressure upon air-cells full of air, and to drive the air hither or thither, according to the amount of muscular force exerted on different portions of the lung-surface.

In addition to this, an elastic body, which has been so far damaged as to have lost the power of complete recoil, must have lost something of the stability of the *whole* of its elastic power; and thus is rendered more susceptible to further damage, through loss of resistance to an amount of tension which, before, it would have been competent to withstand.

In illustration of such changes as I have described, I may refer to cases such as those already cited, in which persons received many years ago some over-stretching of the air-cells, as in the shock of an attack of Whooping-cough in childhood, but suffered no material interference with respiratory power until some superadded affection placed an obstruction in the way of the expiratory tide, and thus taxed the elastic power of the air-cells, and called for forced muscular expiration; or until some change in the occupation of the person called for forcible expiratory acts performed with a closed glottis.

But in an advanced state of Emphysema the walls of the air-cells are found attenuated, their capillaries ruptured and obliterated, and their partitions broken through, so that several cells are thrown into one. Those bronchial tubes, which run among the distended and crowded cells, are subjected to undue pressure from without, and to diminution of their calibre, when muscular expiration is brought to bear upon the Emphysematous lung by which they are surrounded.

II. THE EARLY TREATMENT OF CATARRH.

THE fourth on the list of "conditions, with some or all of which we have to deal in the majority of cases of Winter Cough," is the thickened naso-pulmonary mucous membrane, and narrowing of the passage which it lines.

I have already said that this is the most important of the causes of Emphysema. I have shown that this condition plays the principal part in causing variations in the degree and persistency of the short breathing, whether the Winter Cough is accompanied by Emphysema or not. I have shown that when Emphysema exists in a slight degree, it may be pushed on to any extent by continued obstruction to the expiratory tide by narrowed air-tubes, and that its progress may be stayed by removing the obstruction. I have shown how the Winter Cough itself is dependent, in the majority of instances, upon the condition of the naso-pulmonary mucous membrane. And I have shown the way in which this thickening of the mucous membrane and narrowing of the air passages, takes place. We have seen how various are the causes, both within and without the body, by which flushing of the mucous membrane may be excited. I have shown you that this flushing may be a very transient condition, or may run on into the production of serious and permanent changes in the chest.

The cases which I have laid before you illustrate all the stages of Catarrh of the naso-pulmonary tract, from a mere cold in the head to a severe and abiding disease of the whole bronchial tree. They show the way—the insiduous way—in which Catarrh steals its marches on its victims; how simple in character and short in duration the first attacks may be; how they dispose the mucous membrane to fresh attacks; and how apt each attack is to involve a larger extent of surface than its predecessor; how often it happens that, when once the complaint has reached the finer ramifications of the bronchi, it lurks there still—even when the larger tubes have been restored to temporary health—every fresh attack of Catarrh in the larger passages supplying the lurking enemy with reinforcements, and enabling it to advance from its fastnesses, and to encroach further and further upon the respiratory tract, until at last it needs but a breath of wind upon the lining of the nose or fauces to raise a storm of rebellion throughout the length and breadth of the naso-pulmonary mucous membrane. Thus have we seen how very much that is connected with Winter Cough, in a practical sense, centres itself in Catarrh.

It is to this point, therefore, that I wish to devote most of the time we have left for the consideration of treatment—the treatment of Catarrh, and of those changes in the mucous membrane and in the calibre of the tubes which result from Catarrh.

If we could nip every Catarrh in the bud, what a catalogue of ills we should prevent! And yet this is not such a very difficult thing to do, when we have a chance of trying it. But unfortunately colds are thought so lightly of by patients that they seldom try to stop them till they have become severe, have lasted an unusual time, or have produced some complication. Nevertheless, I believe they would do better in this respect if they had more faith in the possibility of stopping colds; if their doctors would impress upon them more the importance of stopping them; and especially if they knew that *colds can be stopped without lying in bed, staying at home, or in any way interfering with business.*

I shall therefore occupy your time for one minute to tell you my plan of stopping a cold. The plan will not answer if the cold has become thoroughly established; it must be begun directly the first signs of Catarrh show themselves in the nose, eyes, throat or chest,—in fact, before any considerable amount of secretion has taken place. If employed at this stage it is almost infallible. The plan is as follows:—

1. Give five grains of ses-carb. of ammonia, and five minims of liquor of morphia, in an ounce of almond emulsion every three hours. 2. At night give ℥iss. of liq. of acetate of ammonia in a tumbler of cold water, after the patient has got into bed and been covered up with several extra blankets; cold water to be drunk freely during the night should the patient be thirsty. 3. In the morning the extra blankets should be removed, so as to allow the skin to cool down before getting up. 4. Let him get up as usual, and take his usual diet, but continue the ammonia and morphia mixture every four hours. 5. At bed-time the second night give a compound colocynth pill. No more than twelve doses of the mixture from first to last need be taken as a rule; but should the Catarrh seem disposed to come back after leaving off the medicine for a day, another six doses may be taken and another pill. During the treatment the patient should live a little better than usual, and on leaving it off should take an extra glass of wine for a day or two.

As everything depends upon the promptitude of the treatment, persons who are subject to Catarrh, especially if it inclines to the influenza character, should be provided with a prescription for the medicine, and full instructions how to manage themselves *when a cold sets in*. Many old Catarrhal patients of mine, who have been accustomed for several years to stop their colds in this way, have given their medicine the somewhat unprofessional title of the "magic mixture," and would not be without it for the world, so often has it saved them from their old enemy. That, then, is, in my opinion, the best and simplest way of *stopping a cold*. It, in fact, leads to its cure by "resolution." An addition to this plan is needed in persons whose colds seize at once upon the bronchial mucous membrane. Besides the plan of proceeding I have described, they should put ten grains of extract of conium, ʒi. of tincture of benzoin, and ʒss. of sal volatile into a pint of hot water, temperature 170°, and inhale the steam for fifteen minutes at bed-time each night; put a mustard poultice on the front of the chest one night, and between the shoulders the next; and unless the weather is warm, should wear a respirator out of doors till all signs of the cold have quite passed off.

By these simple means, promptly adopted, an at- attack of Catarrh may generally be stopped, and thus all the troublesome and serious effects prevented which follow an established and protracted cold. I would particularly point out that in Epidemic Catarrh or Influenza I have followed this plan of treatment, and it has proved most successful.

If these timely steps have been neglected, and a Catarrh in the naso-pulmonary tract has become fairly established, a different plan of treatment is of course required, which must differ according to the severity of the attack and the part of the mucous membrane principally affected. It may then of course become necessary to confine the patient to the house, to his room, or to his bed, and may involve all the treatment usual for acute Bronchitis, with which you are so well acquainted. With this I will not occupy your time, therefore, further than to impress the great importance of completely curing each attack—of leaving no vestiges behind; we must only be satisfied when, by all methods of testing the respiratory powers, we cannot detect a lingering trace of disease in the naso-pulmonary mucous membrane.

But there is a form of Catarrh, common—I might almost say universal—among the children of the poor, of which we have very little chance of seeing the beginning, although we too often see the end. It begins so early in their little lives that they seem as though they were born to it, and it goes on summer and winter from year to year. It is, of all forms of Catarrh, that which most certainly leads to thickening of the naso-pulmonary mucous membrane and narrowing of the air-tubes. It begins in carelessness and folly, and continues through carelessness and folly. The children are not half-clothed from their tenderest years; the little money at the disposal of the parents is wasted on a few fine clothes, instead of being spent on a sufficient covering of wool next the skin. It is quite sad to see the children brought to this Hospital —martyrs to Catarrh—with tawdry feathers and smart ribbons, but not a scrap of flannel on their wretched little bodies. Besides these defects in dress, they are subject to be taken by their mothers from close, smoky rooms into the cold air, and to be exposed at the corners of draughty streets while their mothers are gossiping late into the evening. While these conditions remain, no medical treatment can be of any avail; and there can be no doubt that it is from this source that a very large number of the cases of narrowed air-passages, chronic Bronchitis and Emphysema, are supplied to Hospitals.

We come, then, to the important question, What plan of treatment is to be pursued in cases of confirmed thickening of the naso-pulmonary mucous membrane, with narrowing of the air passages?

The number of such cases presented at this Hospital is enormous, and I have no hesitation in saying that they are quite susceptible of successful treatment, although the difficulties in the way are often very great. In private practice it is much easier to cure these cases, because proper treatment can be more satisfactorily carried out.

I have already explained that in this state the

mucous membrane is remarkably susceptible to fresh attacks of Catarrh, by each of which its morbid condition is aggravated. The first point in treatment, then, is to provide against such attacks. It is, indeed, the fact that the membrane will, in course of time—a long time, certainly—spontaneously recover its normal condition, if it can be *absolutely* protected from the recurrence of Catarrh. It is in this way that quite wonderful recoveries have been made by fortunate changes in climate. I use the word *fortunate* advisedly, because even the best-selected change is sometimes most unfortunate, in consequence of those vagaries in the seasons to which all climates are subject.

Let me not omit to speak of that very important class of influences which I placed seventh on our list of the things we have to treat—those "general conditions, favoring or producing morbid susceptibility of the mucous membrane of the air-passages."

When treating of the hereditary transmission of a tendency to naso-pulmonary Catarrh, I pointed out that this was due to the hereditary nature of certain diathetic states; and when treating in detail of the properties of mucous membranes, and the various modes in which they are affected, I pointed out that flushing, congestion, irritation, increased secretion, and all the phenomena of Catarrh, may be brought about by the presence in the blood of such impurities as proceed from mal-assimilation, imperfect digestion, rheumatic, gouty, syphilitic, typhoid, rubeoloid, and other poisons; and I pointed out the analogy in this respect between affections of mucous membranes and affections of the skin. I need not, therefore, go over this ground again; but it is in relation to treatment that these facts assume the greatest importance. We may as well expect to cure a skin-disease dependent upon mal-assimilation by external applications alone, to cure an attack of gout by poulticing the great toe, or to remove a syphilitic affection of the eye by the application of simple lotions, as to effectually treat an affection of the naso-pulmonary mucous membrane, dependent upon similar general causes, by remedies directed only to the catarrhal condition of the membrane. We may, indeed, in one case and in the other, produce temporary local amendment by such local and narrow-minded treatment; but we know that the local disease will recur again and again, so long as we neglect the general condition. When I say, then, that the first thing in treatment is to provide against fresh attacks of Catarrh, I must place first on the list of means of such prevention, the treatment of whatever general conditions we can discover, which, acting from within, may favor or produce morbid susceptibility of the mucous membrane of the air-passages.

Everything that we learn from physiology and

pathology, all that we know of etiology, confirmed by our deepest clinical experience, conspires to hold up *the treatment of diathesis as the secret of therapeutic success.*

III. THE AVOIDANCE OF COLDS.

[Having stated that in 72 per cent. of the cases of cough he had examined, he has found the disease aggravated by fresh colds, caught either by

1. Sudden changes of temperature.
2. Fogs and damp air.
3. Draughts of cold air.
4. Getting wet.
5. Wet feet.

He proceeds:]

IT would indeed seem strange if we could not find means of protection against such common-place influences. In truth, there is no deficiency of means of protection against them; and it is because of the very common-place character of these means, and of the influences themselves, that both are so much neglected and undervalued.

But 72 per cent. of the cases of Winter Cough, which I have analyzed, might probably have been prevented by attention to these common-place things. Let us then give a few minutes to their consideration.

1. Sudden changes of temperature.

This is the most difficult to avoid of any on the list.

The occupations and amusements of all classes involve such changes, and we cannot stop these occupations and amusements, even were it desirable to do so. The work-shop, the counting-house, the committee-room, the opera-house, the ball-room, must be warm when the outer air is cold, and changes from one to the other cannot be avoided. But very much could be done to prevent the body from feeling these changes. The first and most important is the complete envelopment of the body and limbs in wool next the skin, thus interposing a bad conductor of heat between the surface of the body and the outer air. It is surprising that even in the present day this simple and common-sense protection is neglected by so large a number of persons, both of the educated and of the uneducated classes. It is not sufficiennt for the purpose in view that a little body-vest should be worn just big enough to cover the thorax and abdomen, leaving all the extremities unprotected. It should be insisted upon by medical men that the arms and legs require to be protected from sudden transitions of temperature, as well as the trunk. In fashionable life the greatest practical difficulty we have to encounter is the question of exposing the necks and shoulders of ladies in evening dress. It is useless to order body-clothing of wool to the throat, and to expect that ladies will give up a fashion which has been followed and thought charming in all countries and all ages. The difficulty is, however, to be got over pretty well. Every lady in evening dress should carry with her, as invariably as she does her pocket handkerchief, a Shetland shawl, or a mantilla of wool or fur, of a size and shape to cover all those parts not protected by woollen underclothing, and it should only be removed while actually within warm rooms, and should be kept at hand to replace on passing through passages, or if the rooms become cold, or if sitting in draughts.

The main source of protection, then, against sudden changes of temperature to the surface of the body, is to be found in a complete covering of wool next the skin. But besides this, a much greater attention than is common should be paid to putting on and taking off complete and efficient *over*-clothing, on going from hot to cold and from cold to hot temperatures. This is particularly neglected by the working classes, and by girls and boys at schools. In fact, schoolmistresses and schoolmasters appear to be peculiarly neglectful and peculiarly ignorant of the grave importance of these matters, as they are of so many others which not less vitally concern the physical welfare of those under their charge.

But when we have adopted all available precautions for avoiding transitions of temperature to the surface of the body, we shall entirely fail in our object of preventing catarrh, unless we also protect the naso-pulmonary mucous membrane itself. But of this I shall speak again by and by.

What I have said with regard to sudden changes of temperature will apply equally to two other causes of fresh colds upon our lists, viz., draughts of cold air, and cold winds. Both are to be deprived of their sting by proper clothing of the skin and mucous orifices.

Getting wet, and wet feet, occupy a very serious place in our list; and there is no doubt that damp and cold applied to the general surface is the most efficient means of producing chill and vital depression, with congestion of the internal organs. It is necessary that cold be combined with moisture to produce this effect. Even if all the clothes on the body are wet, no harm will come so long as they are kept warm; and this suggests the very great value, to all persons liable to exposure to wet, of light waterproof overalls. They may either be put on to keep the under-clothing dry, or if the under-clothing has become wet either by weather or by perspiration, they may be put on to prevent too rapid evaporation and consequent reduction of temperature, especially when the person is about to remain still after getting warm with exercise. In this variable climate, therefore, school-girls, governesses, shop and factory-girls, and all women whose occupations call upon them to brave the weather, ought to carry with them complete waterproof mantles, made as light as possible, but extending from the neck to the ankles, which can be put on or not as required; and boys and men, similarly exposed, should carry waterproof overalls.

These are things easily obtainable in the present day, and within the reach of all classes; so that it only requires that their importance should be sufficiently impressed upon those who need them.

But if wet and cold to the surface of the body is a fruitful source of Catarrh, wet feet—which means wet and cold feet—is a still more prolific source. There is no external influence which so surely produces congestion of the naso-pulmonary mucous membrane as wet and cold to the soles of the feet. There is nothing so universally neglected, and yet there is nothing more easy to avoid. Warm socks, horsehair soles, goloshes, provide efficient protection against wet and cold feet. It does not seem to be half enough understood that, although a shoe or boot may not be wet through, if the sole is damp it will by evaporation most effectually conduct away the heat from the sole of the foot, and therefore ought never to be worn after exercise is over.

I should hardly have ventured to occupy so much of the time of a medical audience with these apparently simple and common-place suggestions, were it not that they are so common-place that their importance is apt to be overlooked.

We have still one item left on our list—viz., Fogs and Damp Air. I have particularly remarked, that although the smoke and other irritating matters constituting fog are unquestionably very injurious, it is the *moisture and cold* of the fog which are the qualities most potent for mischief to the naso-pulmonary mucous tract. There is but one means of depriving a fog or mist of its injurious properties, and that is a respirator; and the same may be said of the changes of temperature, of which I spoke just now; a respirator is the only means of protecting the respiratory passages from the effects of transitions of temperature. It would be difficult to over-estimate the value of efficient respirators in this climate, as a means of protection against naso-pulmonary catarrhs, if persons disposed to these affections would only carry respirators about with them in their pockets, ready to put on if required at a moment's notice.*

I believe that any kind of respirator is better than none; but after experimenting with every kind that

---

* The popular belief that breathing through the nose with the mouth shut is as efficient a means of warming and drying the inspired air as wearing a respirator is a delusion. And when a respirator is worn, inspiration should be performed *entirely through* it.

has been brought out of late years, I am convinced that there is none at all equal to Mr. Jeffrey's metal-wire respirator, or, as he now calls it, "pneumoclime." The "himalene," which he introduced, was also a most excellent instrument; but I object to it on account of the warm scarf in which it is concealed. Although it is quite proper to cover the neck lightly, I am decidedly of opinion that *warm wrappers* round the neck are objectionable; they produce congestion of the nasal and faucial mucous membrane, and thus dispose to the very complaints they are supposed to prevent.

But before leaving this subject of sudden changes of temperature, I must not forget to speak of sleeping-rooms. It is quite astonishing what follies are committed with regard to the temperature of sleeping-rooms. On what possible grounds people justify the sudden transition from a hot sitting-room to a wretchedly cold bed-room, which may not have had a fire in it for weeks or months, it is impossible to say; but it is quite certain that the absurd neglect of proper warming in bed-rooms is a fruitful source of all forms of Catarrh. We cannot too much impress this upon our patients. It may often be almost as necessary for a delicate person to put on a respirator on going up to bed as when going out of doors, unless proper precautions are taken to assimilate the temperature of the sleeping-room with that of the sitting-room.

Such, then, are the principal means by which I would attempt to defeat the fickleness of climate, and to prevent the recurrence of those attacks of catarrh which keep up and aggravate the disease of the mucous membrane. And you will probably have observed that they all assume that the patient suffering from Winter Cough *is to lead an active and an out-of-door life*—not to be confined to his bed-room, or his sitting-room, or even to his house.

This is a point in the treatment which I consider of very great importance.

Shut up your patient month after month, and perhaps winter after winter, in warm rooms, with little exercise, and you need not be surprised if you add fatty degeneration to his emphysematous air-cells; fatty degeneration to his heart, the muscular strength of which is so important in keeping up his pulmonary circulation; biliary congestion to his liver, already disposed to be overcharged with blood; fat to his omentum, to impede the free action of his diaphragm—so essential to his easy respiration; fat to his diaphragm itself; dyspepsià to his digestive organs, the vigor of which is so important in keeping up healthy nutrition in his tissues;—in fact, if you adopt the "shutting-up system," you need not be surprised if, after a dreary hypochondriacal life, your patient should become prematurely old, and die of apoplexy, paralysis or dropsy.

IV. THERAPEUTIC RESOURCES IN COUGH.

1. Medicines introduced by the Stomach. 2. Medicines introduced by Inhalation. 3. Counter-Irritants.

IN the treatment of coughs, the principal agents are:—

1. Medicines introduced into the stomach.
2. Medicines introduced into the air-tubes by inhalation.
3. Counter-irritation.
4. Change of climate.

I do not believe in the possibility of adapting the exact details of treatment to particular cases, without taking into consideration and carefully balancing all the circumstances of each case to an extent which it is impossible to do in lectures and books, or in any other way if the patient is not before us. *I will not pretend, therefore, to direct the exact cases in which this or that remedy or combination of remedies is to be used.* To do this is, in my opinion, very much like the folly sometimes perpetrated by Governments, of issuing from their offices at home orders for the exact mode in which their generals abroad shall conduct their battles. It has always ended in defeat.

Having, therefore, put you in full possession of the principles upon which your treatment is to be based, all that I shall further attempt to do is to call attention to those medicinal armaments at our command which I consider most important in carrying out the principles of treatment indicated in these lectures. The exact disposal of the forces in any given case must be left to the judgment of the man who takes the responsibility of conducting the battle.

1. Of

MEDICINES GIVEN BY WAY OF THE STOMACH,

I would particularly call attention to ses-carbonate of ammonia. When treating the naso-pulmonary mucous membrane we must not look upon ammonia simply as a stimulant; it has a most marked and important action upon the mucous membrane, as it has upon the skin in Erysipelas; probably this is due to its influence on the blood and on the capillaries. It assists more than any one other drug of equal safety in restoring a healthy condition to a mucous membrane affected with catarrhal congestion. It should be combined with other medicines under varying circumstances. Thus, if there is a high state of recently excited injection, tartarated antimony may be given with the ses-carb. of ammonia with the best effect. If there is great irritability of the membrane, morphia in small doses may be given with the ammonia, as I have directed to stop a fresh cold.

With regard to antimony, it is important to bear in mind that it ought never to be continued long. The good it can do is soon done if done at all, and directly it is accomplished the continuance of the medicine does harm. Of ammonia I would also say that, whereas when first administered it acts as a stimulant, it very soon loses its effect and becomes a depressant by its action on the blood. It should never therefore be too long continued. It is better to withdraw it for a time and give it again, than to keep on with it too long at once. With morphia it is necessary to be very watchful that it does not stop secretion, when free secretion is the best means of relief to a congested membrane, and that it does not stop cough, when cough is an indispensable means of clearing tubes choked with secretion. But these are matters with which you are no doubt perfectly familiar.*

---

* Aconite, one of the oldest (see Pliny and Dioscorides) weapons in our armory, has of late years been restored to public confidence by its more careful preparation. It is so powerful a poison that both patients and doctors got to be afraid of it from its uncertain action when clumsily prepared; and thus, instead of being in daily use, it was reserved as a last resource for very important occasions, or used only as an external application. In the treatment of Chest complaints it deserves to stand in the front ranks, and is perfectly safe and manageable in the form of the present Pharmacopœia Tincture. The points to be

Of ipecacuanha and squills I need hardly speak, their effects are so well known—one as a relaxing and soothing promoter of secretion, the other as an irritat-

especially borne in mind are, that it acts very quickly, and therefore, that the danger is concentrated when full doses are given at once although at distant intervals—but that this danger is quite removed by dividing the dose and giving it more frequently; thus, if it is wished to give ♏ xxiv in 24 hours, instead of giving ♏ iv every 4 hours the effect is better obtained and all risk avoided by giving ♏ i every hour, the medicine being stopped at any hour in the twenty-four if it appears to act too powerfully or to disagree. In doses of from ♏ i to ♏ ii every hour, it is of the utmost use in the early stages of naso-pulmonary catarrh. I think its influence is best described by comparing it to the combined good of Ammonia, Antimony, and Morphia, without the special evils of either. But it must be remembered that its use is confined to the stage of active vascular turgescence—that it has a special and important evil of its own, viz., a paralyzing action on the hear —that therefore the heart should be carefully watched during its administration—and that there are cases of weak heart and of mechanical obstruction to the circulation in which it is an unsafe remedy. But still the plan of giving it in minute repeated doses reduces the cases in which it cannot be given to a minimum.

In the treatment of Bronchitis in asthmatic subjects, Aconite may be brought in with the greatest service after Ammonia and Antimony, just as the Antimony has attained its end and begins to depress too much without compensatory good. At this point I am accustomed to substitute the Aconite for the Antimony before withdrawing the Ammonia, and with the best effect.

The properties of this important drug may be well summed up in the following statement of the results of the elaborate and careful experiments and observations of Dr. Fleming:—

ing expectorant, which assists in clearing the membrane of secretions already produced, and stimulates the mucous glands to contract and cease to secrete more. Ipecacuanha therefore may do harm in one way, and squills in another, if too long continued.

Senega irritates the cough. If there is too little cough in proportion to the secretions requiring removal it is an invaluable medicine, but I think its value is restricted to producing this effect. Although when masticated it excites the salivary glands by local irritation, I have never seen reason to believe that as an internal remedy it has this effect on the salivary glands, or that it excites the naso-pulmonary mucous membrane to increased secretion, as some appear to think: on the contrary, its effect is drying and irritating, and therefore senega should *not* be used if secretion is deficient. If the cough is already frequent it does harm.*

"1. That Aconite is a sedative of the cerebro-spinal system, by its direct action upon the nervous matter and on the heart.

"2. That it is a powerful antiphlogistic.

"3. That it is calculated to be of great value in all cases where there is inordinate activity of the circulation.

"4. That it is contra-indicated where there is an obvious mechanical impediment to the passage of the blood, particularly through the heart or lungs."—(Inaugural Address, University of Edinburgh, 1844.)

*Serpentaria (one of the ingredients of Tinct. Cinch. Co.) is a valuable adjunct to our other remedies; it is persistent in its effects as a

Ammoniacum acts by promoting cough, and thus assisting expectoration; and beyond this it has a further effect in stopping secretion, and afterwards producing excitability of the mucous membrane, which may become excessive if not watched.

I consider olibanum a much more valuable drug than ammoniacum, and I am sorry to see it so little used. It has a remarkable effect upon the intestinal mucous membrane, arresting chronic dysentery and diarrhœa, and leaving a soothed condition of the bowel. I have seen the same effect produced upon the mucous membrane of the air-tubes—morbid secre-

---

stimulating diaphoretic, having some resemblance to camphor in its action on the brain skin and mucous membranes. It is of much value in combination with other remedies where there is a "typhoid" or sinking tendency; but for the same reason it is to be avoided where there is nervous or vascular excitement.

Copaiba, from its nauseous flavor and disreputable associations, is less used than it ought to be in affections of the naso-pulmonary mucous membrane, in which its influence is only second to that on the genito-urinary tract. It is in chronic bronchitis and chronic catarrh unattended with fever that it is especially valuable, correcting the tendency to excessive and muco-purulent secretion, and assisting in restoring the mucous membrane to a healthy condition.

Cubebs may be placed in the same category as Copaiba, with this decided advantage—that instead of deranging the digestive organs, it benefits them, and that it is by no means so disagreeable to take. It deserves a very high place among the remedies for naso-pulmonary catarrh, but caution is required in its use, otherwise secretion may be too rapidly stopped and irritation result.

tion checked and altered without that irritation which is caused by ammoniacum. The olibanum may be used both by the stomach and by inhalation. It is probable that its topical effect is the more important

Benzoin is another valuable remedy in restoring a healthy condition to a thickened naso-pulmonary mucous membrane. It may also be used by the stomach or by inhalation.*

Spasmodic contraction of the bronchi indicated, as I have already pointed out, by the high pitch of the *in*spiratory sounds, is a very awkward complication; and if not kept in check, it very much interferes with the restoration of the mucous membrane to a state of health. The most efficient means of relief for this spasm are—1. Smoking Savory & Moore's Datura Tatula. 2. Inhaling the fumes of burning nitre paper. 3. Smoking the "Cigares de *Joy*," sold by Wilcox, of

---

*Chloride of Ammonium is not used as often as is deserves to be, probably in great measure from the difficulty in concealing its nasty taste, but this may now be easily done by the new Pharmacopœial Liquid Extract of Liquorice, which is in other respects also well suited to be given with the drug. The most valuable property of chloride of ammonium in naso-pulmonary affections is its power of rendering the secretions of mucous membranes less viscid and tenacious, at the same time that it is somewhat stimulant or tonic, or at least not depressant in its general effects. These are qualities which make it a most important adjunct to other internal remedies, and in addition to these it certainly soothes and tones the mucous membrane when used as a lozenge or inhalation.

336 Oxford Street. 4. Ten to thirty drop doses of Etherial Tincture of Lobelia Inflata;* and 5. When the spasmodic tendency can be traced to (*a*) gouty,

---

* To the above list of remedies for asthmatic spasm may be added the following, from each of which I have seen good in special cases:—

1. Papier Fruneau (à Nantes).
2. Dr. Palmer's anti-asthmatic papers (Dublin). See Appendix VI.
3. Ozone paper (Huggins, 235 Strand).
4. Papier de Barrel.
5. Martindale's pastilles (New Cavendish street).
6. Cigarettes de Barrel.
7. Cigarettes of Eucalyptus globulus (Savory and Moore).
8. Pariss's Cigarettes Pulmoniques (Bell).
9. Cigarettes d' Espic.
10. Inhalation of ether.
11. Inhalation of camphor.
12. Inhalation of nitrite of amyl (with great caution). "Impure nitrite of amyl inhaled causes violent cough and irritation of the larynx. The pure nitrite has a remarkable effect in causing dilatation of the blood vessels with intense flushing and congestion of the head and face, the skin feeling excessively full and tense. From two to five drops of the nitrite inhaled from lint will give speedy relief in angina pectoris, and in some forms of asthma with pallor of the face. . . . Nitrite of amyl speedily loses its power if kept."—Dr. Thorowgood's "Student's Guide to Materia Medica," p. 142.
13. Cautious inhalation of chloroform.
14. Chloral hydrate in doses of from five to thirty grains. (See note, p. 209.)
15. Indian hemp (Squire) either smoked or given internally. (See note, p. 207.)
16. Belladonna.

(*b*) rheumatic or (*c*) malarious causes, the administration of colchicum, alkalies, quinine, arsenic.

Stramonium, administered by the stomach in quarter grain doses of the extract or twenty drop doses of the tincture, twice or thrice in twenty-four hours, becomes a very useful adjunct to other treatment under these circumstances.

I must not omit to remind you of the facility with which one part of the naso-pulmonary mucous membrane is able to act for the relief of another, as illustrated by the case I related of the gentleman whose spasmodic asthma was often carried off by sneezing, and who suffered from dyspnœa when a discharge of mucus from the nostrils was suddenly stopped (p. 42). We may often take advantage of this property of the mucous membrane with the best effect. Congestive tumidity of the bronchial membrane may be relieved by giving iodide of potassium in sufficient doses to produce the symptoms of a free coryza; and errhines may be used for the same purpose. Iodide of potassium, however, requires to be given with caution; for if it does not succeed in producing a free discharge from the nares and sinuses, it simply inflames the mucous membrane; and this inflammation may run down and aggravate the affection it was intended to relieve. Unless there was a syphilitic taint, I have never been satisfied that iodide of potassium acted beneficially upon a thickened naso-pulmonary mucous membrane in any other way than as a derivative, in the manner described.

I can speak much more satisfactorily of the influence of saline aperients acting as derivatives on a *different* tract of mucous membrane, and free, therefore, from the chance of exciting a tumid or inflamed condition in the diseased one. One of the best forms of saline for this purpose is the Friedrichshall water. There is, however, a very valuable aperient mixture which I often prescribe, containing sulphate and carbonate of magnesia, with a small quantity of iodide of potassium, and a full dose of nitrous ether, which appears to suit some cases better than the Friedrichshall water—producing a freer and more watery discharge, and helping to eliminate gouty and rheumatic poison.

In some cases of considerable and long-standing thickening of the mucous and submucous tissues, it may be necessary to give small doses of bichloride of mercury for a considerable time.* Colchicum, sulphur and arsenic, too, which act so beneficially upon some affections of the skin, are often of great use in treating the respiratory mucous membrane.†

---

* In some cases where the thickening of the mucous membrane is kept up by the frequent occurrence of catarrh, the persistence in small doses of opium will by keeping off these attacks permit the membrane to become restored to its normal condition.

† Where there is a tendency to alternation of affections of the skin

But I am anxious not to leave this part of the subject without mentioning Tonics. Nothing can be more important in the treatment of Winter Cough than to improve the "tone" of the whole system of the patient by every means in our power. Zinc, iron, arsenic and quinine are the most important of such medicines; and I think their value may be taken to stand in the order in which I have enumerated them.* Strychnia

---

and of the naso-pulmonary mucous tract—which may often be observed, especially in the rheumatic, gouty, syphilitic and strumous diatheses—whatever *constitutional* treatment has proved efficacious in the removal of the skin complaint *without exciting its metastasis to internal parts* is likely to assist in the treatment of the pulmonary affection. But it very often happens that in curing a skin complaint asthmatic and bronchitic symptoms are excited, and this shows that the treatment is not efficacious in the sense of attacking the diathetic defect, and the remedies which have this tendency should of course be carefully avoided. In some chronic cases the only way to keep the lungs safe is to maintain the skin disease; and attempt should be made to do this in some part of the body out of sight, and where it will be of least inconvenience to the patient; on the whole, the shins or calves are the best. I have known many cases, especially in the old, in which patients remained free from cough and dyspnœa only so long as they were content to put up with a patch of Eczema on each leg. When this is found to be the case, and the patient is not too advanced in years, the proper plan of proceeding is to make a mild and insidious attack on the diathesis, which should be persevered in steadily till the constitutional condition is radically changed. This is generally best done by mineral waters and baths, and change of climate.

*Tincture of Eucalyptus globulus, introduced since the above was

has been highly recommended in Winter Cough complicated with Emphysema; but as it especially acts upon the muscular system, exciting the muscular fibres to spasmodic contraction, I see no reason to expect any special good effect from it upon diseases of the air-passages and cells; for, in the first place, we wish to diminish and keep off spasmodic muscular contraction of the air-tubes; and, in the case of the overstretched air-cells, they are not muscular but elastic fibres which we wish to strengthen; and there is no reason for supposing that strychnia acts upon elastic tissues. If it does good, it can only be through its effect upon digestion, and perhaps on the muscular walls of the heart. In this way, like many other medicines that might be mentioned, it may do good service; for we must ever bear in mind, that, in order to prevent degeneration of the Emphysematous air-cells, of the walls of the air-tubes, and, especially of the heart, every available means must be employed for keeping up the vigor of the nutritive functions; so shall we not only keep off degeneration of tissue, but promote the healthy repair of the diseased parts.

---

written, has proved a valuable tonic in catarrhal affections, accompanied with a tendency to remittent feverishness, and will sometimes suit when quinine disagrees. This will be worth remembering where there is a complication with spasmodic asthma, and quinine is found to bring on the spasm, as it sometimes will do.

In extreme cases of Winter Cough life may sometimes be saved by the timely and protracted use of ARTIFICIAL RESPIRATION, a means of treatment to which I drew attention in the British Medical Journal, Jan. 30, 1869, in the following "clinical note:"—

"I have long been accustomed to prescribe artificial respiration, as an auxiliary to other treatment, in cases of chest disease accompanied by imperfect aëration of the blood, especially in the young, the weak, and the old. When it has been practicable to get this treatment fairly carried out, I have seen the most pleasing results. But the difficulties are many. It is evident that the Marshall Hall-method is inadmissible, as patients, seriously ill with chest disease, cannot bear to be rolled about. The method by alternating pressure on the sternum, also, cannot be borne, especially when there is heart disease. The Silvester-method, therefore, is the one which I have chiefly employed; but it has these among other disadvantages: 1. It is very seldom that there is room while the patient lies in bed for the arms to be carried back sufficiently. 2. When a patient is sitting up, if weak or suffering from heart disease, raising the arms above the head produces faintness. 3. The operation is very fatiguing and distressing to the patient, as well as to the attendant. The result is that, in the cases in which it is most urgently required, it is quite a chance whether it can be effectively done. It happened that at the time Dr. Bain read his paper "On an Improved Method of Practising Artificial Respiration in Cases of Suspended Animation" (at the Royal Medical and Chirurgical Society, December 8th, 1868), I was attending an exceedingly bad case of capillary bronchitis in an old man, the subject of extensive Emphysema with dilated heart and anasarca. There was the greatest difficulty in preventing him from sinking into hopeless carbonic-acid poisoning. I had been fighting against this for several days by employing an attendant to use artificial respiration by Silvester's method, as often as the patient became more than usually blue and drowsy, but the fight was a most difficult one; the patient was unable to lie down, and the raising the arms, while he sat in an easy chair, fatigued and distressed him so much that it was evident he could not bear it much longer; yet he was quite unable to go on without it. Immediately after hearing Dr. Bain's paper, I visited the patient and altered the plan of proceeding. He sat in a high-backed easy chair. I got behind him upon a stool sufficiently high to give good command over his shoulders from above, and then, grasping them from before with the tips of the fingers towards the arm-pits, dragged him up with sufficient force just to take off his weight from the chair without unseating him, held him so for about three seconds, then let him down, and after a rest of about three seconds lifted him again as before. This was repeated twenty times, and the patient, instead of showing signs of fatigue and distress, said he felt revived, and the improved color of his lips, nails and skin, and the brighter expression of his eyes, plainly told the good effect it had produced upon his blood. He willingly consented to have the operation repeated every two hours. It was easily done by his attendants, and persevered in for four days and nights. The effect was so remarkable, that at the end of that time it was no longer required. I believe the operation saved his life.

I cannot too strongly recommend the adoption of the same plan in all that class of cases to which I have here referred. I would also point out how important it is in the young, the weak and the old, when suffering from diseases impeding respiration, to prevent them from sitting or lying with the weight of the arms pressing upon the chest. The weaker they become, and the more overpowered with carbonic-acid poisoning, the more completely is the weight of the arms and shoulders allowed to sink down upon the thorax, and thus to interfere with respiration. I believe that numbers of lives are lost by neglect of these precautions and of artificial respiration, just at he critical period when the fever and excitement of the congestive stage give place to the lassitude accompanying free exudation.

Under the head of the general conditions predisposing to Winter Cough, I have already intimated the importance of correcting all defects of assimilation and digestion. The internal medicines and diet necessary for this purpose must, of course, differ with the case; and I cannot do more in this place than remind you never to forget this part of the treatment.

I have shown, when analyzing the reported cases under the heads of Colds and Coughs, that of all the cases of Winter Cough, the cough left in summer weather in 45 per cent., and the short breathing was relieved in summer weather in 29 per cent.; and we found this relief to short breathing and to cough was really due to the removal of the irritability and thickening of the naso-pulmonary mucous membrane, under the influence of *warm soothing inhalations in the form of summer air.* These are very important facts to bear in mind with relation to treatment, not only as indicating the importance of wearing respirators, to which I have already referred, but as pointing to

2. THE USE OF INHALATIONS

As a means of restoring the mucous membrane to a healthy condition. Inhalations may consist of fumes, vapors, or atomized fluids.

My own experience of atomized fluids, except as a

means of applying lotions to the nares and fauces, and for stopping pulmonary hæmorrhage,* is not very satisfactory. I object to their use, as a general rule, for affections below the glottis, in which it is necessary to allow the patient to respire during the operation. They have the great disadvantage of conveying too large a quantity of cold moisture into the air-passages, and are thus apt to produce all the evils of severe damp and fog.†

Fumes and vapors have not this objection, and are among the most valuable of our means of acting upon the naso-pulmonary mucous membrane. Nitrate of potass is of great service when used in this way. It appears to act upon the respiratory mucous membrane, when introduced in the form of fumes dispersed through atmospheric air, much as it does on the pharyngeal mucous membrane when applied in the popular form of " sal prunella balls." It refrigerates and causes resolution of that flushed and tumid condition which I described when speaking of catarrh and asthma. It is in this way, I think, that it so signally relieves some cases of spasmodic asthma. You

---

* I have often been able to arrest obstinate hæmoptysis by a spray containing alum, gallic acid, or perchloride of iron, when other remedies had failed.

† This objection is, to a certain extent, removed by more recent spray-producers, in which the spray can be used warm.

will find that the fumes of nitre paper, and those of datura tatula, and stramonium, and the vapors of chloroform and ether, relieve different classes of cases, or the same case in different phases. When the prominent mischief is vascular injection and tumidity, the spasmodic contraction being only excited by the excess of these conditions, nitre fumes give the most relief. When, on the other hand, the nervous element is most prominent, the tendency to spasmodic contraction being so great that it is set up by a comparatively slight amount of flushing of the mucous membrane, datura tatula, stramonium, chloroform, and the like, are the most potent remedies. The influence of the fumes of nitre upon the bronchial spasm is only a secondary effect, its most important influence being that it removes congestion and tumidity of the naso-pulmonary mucous membrane.

Fumes of carbolic acid are extremely useful as a means of diminishing excessive purulent expectoration, and of removing the fœtor of discharges from the lungs of those in whom disintegration of lung substance is taking place.

The only satisfactory apparatus 'for this purpose, with which I am acquainted, is the "carbolic acid vaporizer," introduced at my suggestion, by Messrs. Savory and Moore. This should be used three or four times a day in the patient's apartment, and especially the last thing at night, so that the atmosphere of the sleeping room remains impregnated with carbolic acid throughout the night.

The other materials which I use as fumes are gum benzoin, gum olibanum, and camphor.

For the inhalation of vapors I am accustomed to order Nelson's inhaler, to be used *without the sponge;* a pint of hot water being put into it, the temperature of which should not exceed 170° when inhalation begins.* The patient should inhale for about fifteen minutes at a time, and should not go out of doors the same day that the inhalation is used, without a respirator. It is often necessary, therefore, to restrict the use of the inhalation to the evening, and thus we very much limit its usefulness. Much better results would be obtained by all these topical applications if we could secure their more frequent repetition—if, in fact, we could apply them to the naso-pulmonary mucous membrane as persistently as we apply a lotion to a skin disease; and the object should be to do this to the fullest extent possible under the circumstances.

The materials which I principally use to medicate the vapor of water are—

1. Compound tincture of benzoin.

---

*A more satisfactory inhaler for general use has recently been made at my suggestion by Messrs. Maw. It is called the "Economical Inhaler."

2. Tincture of iodine.
3. Carbolic acid.
4. Creasote.
5. Spirit of camphor.
6. Spirit of chloroform.
7. Spirit of ether.
8. Juice of conium.
9. Chloride of ammonium fumes, produced by adding a few drops of hydrochloric acid after liq. of ammonia has been mixed with the hot water.
10. Liq. of ammonia, or aromatic spirit of ammonia.
11. Tincture of myrrh.
12. Tincture of lobelia.
13. Tincture of stramonium.
14. Acetic acid.
15. Turpentine.* For the purpose of inhalation, Messrs. Hanbury, of Plough-court, have introduced several specimens of turpentine, the odor of which is far preferable to that of the turpentine in common use.

There is no class of complaints in which

3. COUNTER-IRRITATION

gives such unquestionable and unqualified relief as in

---

* A very valuable way of using tuprentine in chronic bronchitis is to evaporate two or three tablespoonfuls in the patient's sleeping room during the night. This is easily and safely done by putting it into the upper portion of "Clark's Pyramid Food Warmer," taking care to put water in the lower receptacle.

affections of the respiratory tract of mucous membrane. The relief to the oppressive dyspnœa and to the irritability of tubes long narrowed by thickened lining, which speedily occurs under the influence of decided counter-irritation, is delightful to witness*. In severe and chronic cases, I am in the habit of ordering at the commencement of treatment three blisters, one for the front of the chest and one for each side, between the scapula and the breast, to be used in succession; each blister to be allowed nearly to heal before the next is applied. The application of the first I *insist* upon, the other two I leave to the patients' judgment to apply or not, as they choose, after having found what relief is obtained by the first. It is very rare indeed to find that they fail to apply all three—much more often they are so pleased with the effect as to volunteer to put on more if necessary. No form of counter-irritation is at all equal to a well-managed blister. Great caution should be taken not to have the blisters on long enough to produce deep sores troublesome to heal. They should be quickly removed, and a warm linseed poultice applied before the vesicles are cut.

---

* In the acute stage dry-cupping by the elastic bottle is the most convenient, rapid, and efficient mode of relieving congestion, and it has the advantage of not interfering with the subsequent use of blisters or other counter-irritants.

When the laryngeal and tracheal membrane is that most affected, little strips of vesication may be produced along the course of the trachea and at the sides of the neck, in the rear of the thyroid cartilage. It is better not to apply blisters over the front of the larynx. When the posterior nares and pharynx are most affected, the back of the ears or back of the neck are the best seats for external applications.

After the more decided effect has been obtained by blisters, iodine liniment and camphorated ammonia and turpentine liniments may be employed with advantage; but if the case is severe or of old standing, and the patient will allow it, never fail to start with the blister. One caution, however, is needed, viz., not to blister while the skin is hot and dry, and the patient suffering from active inflammatory excitement.*

In some stages of congestion of mucous membrane, especially where the tumidity is great, and there is much reluctance to secrete freely, hot fomentations applied as near the affected part as possible give great relief, and, by promoting secretion, put the membrane in a better position to be benefited by other treatment.

---

*In the rage of the present day for novelties, the old-fashioned pitch plasters are apt to be despised, but great comfort and benefit may be obtained from a good-sized pitch plaster over the front of the chest and between the blade bones, renewed from time to time throughout the winter.

It is surprising how much good may be done in a short time by repeated hot fomentations and poultices to the back and sides of the neck in affections of the nasal, pharyngeal, and laryngeal mucous membrane.

By a combination of such medicaments, appliances, and hygienic regulations as I have now briefly mentioned, we may hope to remove the susceptibility of the naso-pulmonary mucous membrane, and to reduce it to its normal thickness, so that the air-tubes may be restored to their natural calibre, and the obstructions to the expiratory tide removed. The importance of accomplishing these ends, you will perfectly understand from what we have already seen respecting the causation of Emphysema, and the circumstances which aggravate it when it exists and give poignancy to all the sorrows which it inflicts.

In prescribing

IV. CLIMATIC TREATMENT

For Winter Cough, the five clinical groups into which I divided our cases in the beginning of these Lectures, will again claim the first place in your considerations.

You will recollect that these groups consist for the most part of various combinations of Bronchitis and Emphysema, and that in investigating the causes of Winter Cough, we found Naso-pulmonary Catarrh to be the most potent and most frequent of all.

It is upon Naso-pulmonary Catarrh that climate has the most important and satisfactory influence; and whether we are treating a case in which it is the sole morbid condition, or one in which it has led to Emphysema, we shall always have to give it the first place in forming our decision as to the kind of climate to recommend. But this affection, as we have now seen, passes through different stages, and is dependent upon different diathetic causes, all of which must be considered in prescribing climatic treatment.

For practical convenience we may consider Naso-pulmonary Catarrh under the following heads:—

1. Morbid sensitiveness of the mucous tract.

2. Inflammation of the mucous membrane, with deficient secretion.

3. Inflammation of the mucous membrane, with excessive secretion.

4. Chronic thickening of the mucous membrane, the result of the repetition of the second and third of the above conditions.

5. Chronic thickening, the result of one severe, neglected and protracted attack of Naso-pulmonary Catarrh.

6. Disintegration of lung-substance, the result of repeated attacks of neglected catarrhal congestion, or of one such attack occurring under unfavorable diathetic or other conditions.

For practical purposes, also, it will be convenient to consider the Emphysematous complications of Naso-pulmonary Catarrh, under the following heads:—

1. Emphysema, due to accidental overstraining of the air-cells, independent of catarrh.

2. Emphysema, the result of repeated temporary obstructions of the air passages.

3. Emphysema, the result of permanent obstruction of the air passages.

4. Emphysema, the result of recent and still present obstruction of the air passages, by an acute attack of Naso-pulmonary Catarrh.

5. Emphysema of great extent, and of old standing, and with all the indications of broken down air-cells and tissue degeneration.

The diathetic states which will principally call for consideration in climatic treatment are the Rheumatic, the Gouty, the Strumous, the Tuberculous; and the complication of a tendency to spasmodic asthma will always be an awkward one to deal with, and one which will not permit of neglect.

The principal headings into which I have divided Naso-pulmonary Catarrh and Emphysema for climatic consideration involve these prominent questions:—

1. Whether the naso-pulmonary disease is accompanied with great susceptibility to repetition, or is only the result of some unfortunate combination of

circumstances not disposed to recur. In the latter case, of course, we shall be free to select a climate best suited to the removal of the thickened mucous membrane, with less regard to its anti-catarrhal qualities; that is to say, we may place tonicity and warmth above equability.

2. Whether the naso-pulmonary affection is complicated with a relaxed condition of the membrane, and perhaps with dilated tubes, accompanied by excessive secretion, or with an irritable inflammatory condition with deficient secretion. In the first case the summer warmth of climate must be combined with dryness, in the latter with moisture; that is to say, we shall require for the one warm and bracing climates, and for the other sedative climates.

3. Whether the Emphysema is accompanied with a constant susceptibility to fresh attacks of catarrh; or is the result of some past attack, without the tendency to its recurrence being particularly strong, although its effects upon the mucous tract have become permanent. In the former case the anti-catarrhal qualities must engage our first attention in the selection of a fitting climate; in the latter we may allow the consideration of removal of the chronic thickening of mucous membrane to rule our choice, with less regard to the dangers of recurrent catarrh; and thus again we shall have to decide between the warm and bracing and the

sedative types of climate. But we may also introduce the consideration of selecting a climate especially on the grounds of the impregnation of the atmosphere with such matters as turpentine, iodine, or sulphur.

4. Whether the Emphysema, having resulted from accidental straining of the air-cells independent of catarrh, as for example, when it has resulted from over-lifting, over-running, or from the violent overstrain of a long past fit of convulsive coughing, such as whooping-cough. These cases resemble the herniæ of the surgeon's practice—the mischief is done, and the cause of the damage has past. The treatment must consist in informing the patient of the various accidents and influences he is specially to avoid as likely to renew or aggravate his complaint, and in adopting such direct treatment as may help to cure the local damage he has received. We may therefore throw all our consideration as to climate into the question of selecting that best suited to the Emphysema *per se*, or simply that which the general health of the patient requires, independent of his local disease.

And this brings us to the question which I propose to treat here, viz.:

5. The necessity for considering the diathesis of the patient when prescribing climatic treatment for his Winter Cough. It is a matter of great importance; for whether his complaint has been brought about

through a rheumatic, gouty, syphilitic, strumous, or other constitutional taint, the other effects of climate will be most materially influenced by those which it may have upon these general conditions of the system, where the taint is strong and unquestionable in its character. I advise you to give it the foremost place in guiding your choice of climate; for you will do but little towards the radical cure of the local complaint till you have removed its constitutional cause.

The tuberculous diathesis adds frightfully to the perils of all catarrhal affections, and will claim our attention on this account rather than as a *cause* of Winter Cough. When the constitutional tendency to consumption is at all marked, we shall have to be constantly on our guard against the occurrence of local congestions and inflammations during attacks of catarrh, for *whatever increases the vascularity of an internal organ in the tuberculous diathesis involves the risk of tuberculization of the affected part.* And again, when we discover physical signs of that lung-disintegration, which is one of the serious effects of repeated and neglected catarrhal attacks, grave as its importance must be allowed to be under the most favorable diathetic conditions, the co-existence of a tuberculous predisposition at once invests it with all the horrors of advancing consumption of the lungs, and will necessitate a choice of climate entirely ruled by this consideration. To prevent the Winter Cough running on into consumption will be the first consideration in the climatic as well as in every other form of our treatment of the case. But the exact quality of the climate to be selected for the tuberculous will differ so materially according to the circumstances of each case, that I must refer you to my works on consumption* for these details, as they would lead me beyond the limits of this lecture if attempted to discuss them here.

With regard to the influence which a tendency in your patient to spasmodic asthma ought to have upon your councils, I will only say, at present, that you will be frustrated in all your endeavors to help him if you do not make a fortunate selection of climate in this respect; and I may add for your comfort that it is the most fickle difficulty of all! For you will never be able to be sure what climate will suit the asthma till you have tried it, and it will as often as not be the very one which appears the least suited to all the other conditions of your patient.

A word must now be said about the cough *per se*. We have seen that frequent cough, and especially convulsive cough, is one of the causes—and a very potent cause—of Emphysema. It will be very impor-

---

* "On the True First Stage of Consumption," p. 47.—Churchill
"On Tuberculosis," &c., pp. 44, 45.—Churchill.

tant, therefore, to allay the severity of the cough by some direct means while the more soundly curative treatment is being pursued.

This may be most effectually done by the sedatives and anti-spasmodics of which I have already spoken. But I may here especially mention three very valuable remedies—Indian hemp, lobelia, and bromide of ammonium. They have all the great advantage of acting speedily and decidedly. If they do not have a decided and speedy effect—if they do not, as it is popularly expressed, " act like a charm"—they do no good, and it is better to throw them aside. At least, that is my experience.

Indian hemp is rather a troublesome remedy, from the difficulty of getting it of a reliable and equal strength; but Mr. Squire has paid more attention to this point than any other chemist, and, therefore, I now always prescribe his preparations of Indian hemp. Indian hemp is most likely to do good when the cough is accompanied by tendency to fitful bronchial spasm, or when it occurs, not in long convulsive fits at distant intervals, but in a never-ceasing wearing and tearing bark.

Etherial tincture of lobelia has great power in relieving the asthmatic spasm in a certain number of cases, and in stopping those convulsive fits of coughing which occur almost like Whooping-cough, with distinct intermissions. It increases secretion. It often acts quite magically upon those distressing and exhausting attacks of convulsive cough which are, of all others, the most likely to produce Emphysema. It is in this same description of cough, especially when combined with hysterical excitement, that bromide of ammonium often proves useful.*

I must just mention bismuth as another direct means of stopping cough, though it acts by an indirect route. If you find that the cough always comes on when the stomach is empty, bismuth will stop it after everything else has failed—that is, everything except cod-liver oil and food—and a patient cannot always be eating, or sipping oil for the sake of keeping off cough. There can be no doubt that the bismuth in these cases acts through the gastric fibres of the pneumogastric nerve. Hydrocyanic acid will

---

*Chloral.—To the list of remedies having a rapid action on the cough must now be added Chloral Hydrate. In doses of 5 to 10 grains, dissolved in water and flavored with Liquid Extract of Liquorice, it is a most potent remedy in some cases of irritable and paroxysmal cough; but unless it acts beneficially at once, it should be discontinued.

Conium.—I have referred to the use of Conium as an inhalation, but not to its internal use. It is, however, a most valuable sedative for some irritable coughs, and may be given either in the form of juice, extract, or tincture. It appears to act as a sedative to the spinal cord.

often have a similar effect, but the bismuth is a safer and more persistent remedy.

Digitalis is of great use when the cough is kept up or excited by a too rapid and fitful action of the heart. Digitalis steadies the heart.*

The less important, but still useful means of allaying cough are very numerous; but I will not detain you by enumerating them, as they are most of them familiar to us all.

---

* Bimeconate of Morphia (Squire).—When an opiate is unavoidable, the Bimeconate of Morphia should be the form selected. It disagrees much less with the digestion than opium *per se*, constipates less than other forms of Morphia, and is less inclined to produce headache and other unpleasant symptoms. I may here mention that I have ascertained by repeated experiments, that the use of Pancreatine after meals will keep up good digestion in spite of the paralyzing effect of opiates. This is a most important clinical fact, removing the greatest of the difficulties which lie in our way when we find it necessary to continue the use of opiates. I have seen 3 and 4 drachms of Liquor of Bimeconate of Morphia taken daily, without any defect of digestion occurring so long as a dose of Pancreatine was taken with each meal, in a patient, who, without this aid, vomited all solid food undigested whenever the Bimeconate or any other opiate was taken.

V. THE TREATMENT OF POST-NASAL CATARRH.

I MUST not forget to mention the treatment of Post-nasal catarrh. Although at first sight a very trifling complaint, Post-nasal catarrh is unquestionably very troublesome to cure, and is very apt to return. The difficulty is mainly due (1) to the awkwardness of applying topical remedies to the parts principally affected; (2) to the almost invariable existence of a diathetic cause; and (3) to the length of time which the complaint has usually existed before the patient comes under treatment. As in all affections of the naso-pulmonary mucous membrane, the first point is to make out the nature of the existing *morbid constitutional state*, and to apply appropriate diathetic treatment for its removal. But, unfortunately, this alone will not be sufficient; for the local affection will seldom yield without some topical aplications; and it is in settling the form of this local treatment that I have found the greatest trouble. After trying a great number of applications in the form of spray, injection, gargle, lotion, inhalation, snuff, and lozenge, I have come to the conclusion, that the best for the majority of cases is the combination of:—

1. A medicated injection.
2. A medicated snuff.
3. A medicated lozenge.
4. A rubefacient liniment.

The injection consisting of borax, ʒj.; glycerine of carbolic acid, ʒij.; bicarb. of soda, ʒj. to half a pint of warm water.*

The snuff consisting of camphor, tannic acid, white sugar, and high-dried Welsh snuff, of each ʒj.

The lozenge consisting of camphor, gr. ij.; tannic acid, gr. $\frac{1}{2}$; hydrochlorate of morphia, gr. $\frac{1}{36}$; white sugar, gr. x.; acacia gum, gr. ij.

The best rubefacient is pure compound camphor liniment.

The injection should be used night and morning. From three to four lozenges should be taken each day, one of which should be taken at bed-time, and one on waking in the morning. The snuff should be used once in the evening, and once or twice in the day, and it is best applied by means of a little elastic tube, one end of which is charged with snuff, and pushed into the nostril, the other end being put into the mouth,

---

* This may be varied by substituting for the Borax, Chloride of Ammonium, or Chlorate of Potass or Condy's fluid, in certain cases.

and the snuff blown up the nose with a slight puff.*
The liniment should be well rubbed behind the ears
and at the back of the neck twice or thrice a day.

I ought to add that the snuff must be discontinued
should a fresh attack of nasal-catarrh happen to set in,
but should be resumed on the subsidence of inflammatory symptoms.

---

* The lozenges, snuff, and tube are kept ready prepared by Messrs. Bell, Oxford Street, Messrs. Savory and Moore, New Bond Street, and other leading chemists.

VI. THE MANAGEMENT OF CONSUMPTION—REST TO THE LUNGS.

IN November, 1872, at my request, Mr. Bigg made the first "lung-splint," and I asked him to call it at once by this name, as conveying an unmistakable explanation of its objects. Before this, I had been accustomed to procure rest for portions of lung by other devices, principally by keeping the arm of the affected side flexed upon the walls of the chest, so as to restrain expansion by its weight and by the absence of muscular action; and it was the difficulty of sufficiently localizing the pressure by these means to suit special cases that led me to suggest the "lung-splint;" but, whether it be partial rest of the whole of one lung, or more complete rest of a portion of one or both lungs that is desired, the greatest caution is necessary; because whatever local means secure rest to one part of the lungs, throw extra work upon the other parts, and may, therefore, easily do more harm than good.

For this reason, I have always used the greatest circumspection in selecting cases for this kind of treatment, and I trust that, if any one is led to follow my example in one part of this treatment, he will most scrupulously do so in the other. With this precaution, nothing can be more satisfactory or more common sense, in the treatment of lung-disease, than the use of lung-splints, bandages, and the like; whereas without it nothing can be more foolish.

The rules for the cautious application of *localized rest* in lung-disease which I recommend, as dictated by a consideration of the nature of tuberculosis, and justified by the results of my own practice, are as follows:

"1. If one lung, or a portion of one lung, or a portion of each lung, has become diseased, under circumstances which make it certain that there is no constitutional cause of lung-disease, then it is safe to secure localized rest for the diseased part, and to throw the extra work upon the sound parts; but even then it is necessary to be cautious that the extent of lung so rested is not too large in proportion to the extent of sound lung upon which the extra work is thrown. If there is any question about this, rest of the whole body must be secured in addition to the localized rest of lung, so as to save the sound lung from as much work as possible.

"2. If there is a constitutional cause of lung-disease, but only a small area of lung at present suffering, and that in the upper lobes, while there is a capacious chest with large areas of lung in the lower portions quite sound and insufficiently used, then it is safe to secure localized rest for both upper lobes, and to make the lower portions do a fairer proportion of the work; but even under these circumstances, the respiration should be kept at as low a point as practicable. A case illustrative of this rule has just occurred to me. A fine young man, with a very capacious thorax, who has practiced all sorts of gymnastic exercises with his arms while restricting the lower parts of the chest by dress, has thus acquired a habit of breathing almost entirely with the upper portions of lung. He has a tuberculous family history; and, after foolish overtraining, by which he reduced his flesh considerably, he overtaxed his lungs in a race, and he has since become the subject of partial consolidation of the apices and recurrent hæmoptysis. Finding that he has large tracts of scarcely utilized lung at the lower parts of the chest, I have not hesitated to get Mr. Bigg to apply mechanical restraint, by means of lung-splints, to both upper lobes; but I have, at the same time, secured rest for the whole lungs by sending the patient on a long sea-voyage to a warm climate, under careful watching against over-exercise.

"3. If a portion of lung has become disintegrated, under the influence of constitutional causes, and remains obstinately unhealed after all constitutional symptoms have been arrested, and, for some time past, no other portions of lung have shown a tendency to yield, then I think it is quite safe to secure localized rest for the disintegrated portion, so as to give it a fairer chance of healing; while an amount of air and exercise may be allowed to the patient, for the purpose of improving his reparative powers, which could not be permitted while the damaged lung was exposed to the same amount of action as the sound parts. But even here the utmost caution is required not to carry the exercise beyond a very limited amount.

"4. If the constitutional tendency to lung-disease —'the abnormal physiological state'—is strong, and signs of impending mischief in the lungs are scattered, no localized rest should be attempted, but every means should be brought to bear upon the important object of maintaining respiration at its lowest point, consistent with life and nutrition, until the constitutional tendency has become passive and the local symptoms have been removed.

"In conclusion, to prevent misapprehension on so vital a point, let me remind my readers that, in urging 'the importance of rest in consumption,' I am referring to cases in which the lungs are already damaged, or in which the constitutional disease has declared itself in sufficient force to render tuberculization imminent. 'If the symptoms are only what is commonly called premonitory, that is, if they are those of commencing tuberculosis, and no reason or sign is discoverable which justifies the suspicion that tuberculization has commenced; if a sufficiency of fat remains without calling upon the albuminoid tissues, the principles treatment are of quite opposite to those above detailed.'

"It must be admitted that the proper regulation of this matter is one of the greatest trials of the astuteness of the physician, and it is almost impossible, unless he can make the patient and his friends comprehend its meaning and importance. But not less does it test the skill and judgment of the physician to decide upon the moment when restrictions upon fresh air and exercise ought to be removed. The argument so often used when a patient appears to be 'doing well,' that 'it is the best to let well alone,' may be fatal if applied to this case. The very fact that he is 'doing well' may be the sign that he must not be 'let alone;' that he is now in a state in which it is safe to make a call upon his mechanical force, to accelerate histogenesis, to supply fresh oxygen—in a word, to set about the restoration of active nutrition. And then, again, how scrupulously these new tasks should be set; how carefully watched in their effects, lest even now they cannot be continued with safety! On the first sign of their being badly borne, they should be moderated or promptly stopped."

PANCREATIC EMULSION IN CONSUMPTION.

DR. DOBELL was the first to introduce to the profession "Pancreatic Emulsion" and "Pancreatine" as therapeutic agents in Consumption. He was led to employ them by having his attention directed to the fact that many consumptive patients dislike fat. The results of a careful examination of a series of cases which came under his notice at the Royal Hospital for Diseases of the Chest, proved that this dislike for fat is common to the great majority.

With a view of testing by direct experiment, whether this dislike for fat was due to some abnormal condition of the pancreatic secretion, Dr. Dobell determined to treat a series of cases of consumption with the pancreatic juice of the pig. After many tedious experiments, an emulsion of beef-fat, with the pancreatic juice, was adopted as the most eligible preparation. This was supplied to the patients, who were ordered to take it stirred in milk. The emulsion could be mixed readily with the milk, and, in the proportion of half an ounce to a breakfast-cupful of milk, formed a drink that was not unpleasant. Twenty-four patients of the thirty-three treated with the emulsion were discharged after eight weeks, in an improved condition with respect to their general symptoms. The emulsion disagreed with three patients only, whereas cod-liver oil disagreed with eleven out of the twenty-four to whom it was administered. A second series of cases were afterwards treated with the pancreatic emulsion of fat, or with pancreatic emulsion of lard oil, and similar satisfactory results were obtained.

In 1867, Dr. Dobell delivered, at the Royal Hospital for Diseases of the Chest, a course of lectures on " The True First Stage of Consumption," from which the following explanation of the remedial properties of Pancreatic Emulsion is quoted:]

My experience of the action of pancreatic emulsion is now so large, and my observations have been so cautiously and doubtingly made, that I dare to speak with a confidence which I trust may be distinguished from dogmatism. Pancreative emulsion of solid fat is a natural substitute for the inactive or perverted pancreatic function. It supplies the lacteal system with solid fat in a condition fit for absorption, fit for transmission through the lymphatic glands, fit for combustion in the pulmonary blood, for the protection of tissues, for histogenesis, and for general utilization throughout the organism. By an artificial expedient we supply the missing elements of normal nutrition in a natural form. Thus time is gained, the imminence of tuberculization is removed, and the means for the restoration of the normal function of the pancreas, by which alone a true cure is to be effected, can be adopted at leisure and in safety, under conditions favorable to success.

Cod-liver oil, even when it agrees and passes into the blood, does not completely represent the solid fats of the natural food, and therefore cannot permanently take their place. As a temporary substitute for natural fats introduced by the natural route, it answers admirably; but sooner or later, in some cases very soon indeed, the portal system becomes choked, and refuses to absorb more oil; the oil disagrees with the stomach, it rises, it spoils the appetite, and thus not only ceases to do good, but does positive harm by preventing the patient from taking as much food as the stomach might otherwise call for and digest. None of these disadvantages occur with well-made pancreatic emulsion of solid fat. The consequence is, that an artificial supply of natural fat by the natural route can be kept up for an indefinite time, if required, while the appetite is usually improved and the digestion also; and, at the same time, a very large quantity of amylaceous food is rapidly converted into dextrine and sugar by the pancreatic action of the emulsion, and thus a most important assistance in the economy of fat is given by the increased supply of carbon from the carbo-hydrates, at the same time that fat is being thrown into the blood by the emulsion."

PART III.

SOME PRINCIPLES OF DIET IN DISEASE.*

Principles upon which to arrange the Diet of Disease—Rules for carrying out these Principles—Examples of Diets for Consumption and for Diabetes.

THE first and leading principle of diet in disease is, to provide for the maintenance of healthy nutrition under the peculiar alterations of circumstances attendant upon disease. In other words, the diet of disease should be as nearly that of health as the altered conditions of the nutritive functions, and the altered conditions of the patient's life will allow; the grand object being to keep up healthy nutrition of the whole organism.

The true appreciation of this first principle, in its various bearings, will save us from running into either of those extremes which at different times have disfigured medical practice. We shall not be led to starve our patients upon water-gruel when they are craving for natural food; or to stuff them with beef-steaks and

---

* Substance of a lecture delivered by the Author at the Royal Hospital for Diseases of the Chest, 1865.

porter when every instinct of their nature rebels against it. In fact, it may be taken as a very safe rule, that it is better cautiously to supply a patient with the kind of food that his stomach calls for, than to deny it to him without an unquestionably good reason for so doing.

This may seem very much like letting a patient eat and drink what he pleases. But that is not at all what I intend to recommend. All I mean is this, and I wish particularly to impress it, that if we intend to interfere in the subject of diet, we must take care that we thoroughly understand what we are about; and in order to do this it is necessary to keep well up in the following subjects :—

1. The physiology of healthy nutrition.

2. The composition of food, and the essentials of a normal diet.

3. The physiology of disease.

If we keep these matters well before the mind, and at the same time keep our wits about us in watching the case under treatment, it is surprising what an immense deal of good may be done by interfering with the diet; but not otherwise.

We start, then, with this as our first principle—never to be lost sight of—that healthy nutrition is to be maintained, if possible, under all circumstances. And we assume that to do this in a " healthy adult man of average stature taking moderate exercise," the essentials of a normal diet must be supplied.

It must be borne in mind, that the proportions and quantities of the different elements of this normal diet are arranged to meet the requirements of the different functions of the organism when in a healthy state of activity; and it therefore follows, that if the activity of any of these functions is altered, the requirements will be altered; and hence, the second general principle is this:—To alter the quantities and proportions of the elements of a normal diet to correspond with any alterations in the conditions of life. Thus, when a man is overtaken by sickness, and confined to his room or bed, the adult man taking moderate exercise becomes *an adult man taking no exercise;* and the ingredients of his diet which were proportioned to his moderate exercise must now be proportioned to his no exercise; and other alterations must be made in like manner, to correspond with other altered circumstances, in addition to any that may be specially required by the nature of his disease.

But I must remind you that, even when a man is confined to his bed, and precluded from taking any kind of exercise, he is still necessarily undergoing a considerable amount of muscular exertion, which must be provided for in his diet. For example, so long as life remains, such all-important muscles as those of respiration, and the heart itself, continue to act, and to require that their healthy nutrition shall be provided for by a suppy of plastic materials in the food.

We come next to the long list of alterations of function which may be involved in the term "Sickness." And the third principle is: To alter the forms, quantities, and proportions of the elements of a normal diet, to meet the altered relations in the activity and condition of organs consequent upon disease. It is evident that, in order to carry out our first principle of maintaining healthy nutrition under all circumstances, it may be necessary, under some conditions, to reduce the quantity of every element of diet; and also, under some circumstances, to alter the proportions of the different elements. This we see demonstrated in some of the lower animals by the phenomena of hybernation. When an animal gives itself up to its winter sleep, every vital function is reduced to its lowest degree of activity, and the animal is able to maintain healthy nutrition for a long period without taking any food at all; but as respiration has to be kept up more actively than the rest of the functions, a special store of carbon for this purpose is laid up beforehand in the body.

Now, supposing a man to suffer from any state of disease which should place him in the position, as regards his functions, of an animal during hybernation, it is clear that, while his whole diet must be reduced to a very low scale, the heat-giving elements must be supplied in quantities out of the normal proportions as compared with the rest; because no supply of carbon is stored up in preparation for his illness, as it is in the hybernating animal in preparation for its sleep.

We see conditions in many respects similar to these in some stages of fevers, in which absorption, nutrition, and every vital function is at its lowest point consistent with life, respiration being the only one sufficiently active to call for any considerable supply of food. But here, of course, we must not lose sight of an element in the case not present in hybernation—viz., the existence of a poison, which by some means, natural or artificial, has to be eliminated or destroyed, and which may be keeping some functions in activity, the requirements of which must be met. The precision with which we are able to do this in any given case, will depend upon the correctness of our knowledge of the nature of the poison, and of the organs concerned in the restorative process. Here, no doubt, we are often obliged to act in the dark, and to supply many ingredients which may not be needed, in the hope of furnishing among them that which is required, but which our ignorance prevents us from identifying. And we had far better, whenever knowledge is at fault, act in this safe manner and supply much that may be useless rather than run the risk of withholding that which may be essential to life. But, in the majority of cases, our knowledge will be sufficient for the emergency, if we keep in mind the general principles of action.

The fourth principle is this:—To obtain rest for every organ while it is suffering under active disease, by removing from the diet such elements as increase its functions. These are conditions which it is not always easy to fulfil without deviating from our first principle. For example—in the case of diseased kidney—the healthy nutrition of this organ requires a supply of albuminoid materials, while its function is increased by any surplus of these materials in the organism; and when its function is interrupted by disease, a proportion of albuminoids in the diet, necessary to the healthy nutrition of the organism generally, will be tantamount to an excess as regards the function of the kidney, and the accumulation of retained excretory matters will press injuriously upon the affected organ. In such a case other medical aids than diet must be brought to bear; and while the albuminoids in the food are reduced as low as is consistent with healthy nutrition, some auxiliary organs which are not damaged must be stimulated for the time, to save the diseased part from undue pressure upon its functions.

A simpler, but still important principle, may be stated as the fifth, viz.:—In all alterations of diet, to avoid any unnecessary reduction in the number and variety of the forms in which food is allowed to be taken. This becomes especially necessary to be borne in mind when dieting the dyspeptic, who are often still engaged in the active avocations of business and of society while under medical treatment. To treat such cases by cutting off from the daily bill of fare first one article and then another, till the food consists of only two or three permitted forms, is to destroy the appetite and the digestive powers by monotony of diet, and to depress the spirits of the patient by a constant series of petty denials. This plan of dieting can only be regarded as the resource of ignorance; because an enlightened view of the case will discover some particular defect in the function of digestion or assimilation which will at once indicate the form or element of the food which is to be avoided; and thus it will be only necessary to cut off those articles which specially represent this element, or simply to alter the forms in which they are presented to the stomach.

The sixth principle is also of great importance, viz:—When it is necessary to remove from the food any of the essentials of a normal diet, to aim at selecting that which will answer the desired end with the least danger to the nutrition of the vital organs. For exsample, if it is necessary for any special purpose to diminish the heat-giving elements of the diet, it is safer to remove the carbo-hydrates than the hydro-carbons, because the latter not only supply carbon for the evolution of force, but are essential to the nutrition of the nervous system, and of the albuminoid tissues generally.

The seventh and last principle which I shall give in this Lecture is of very general application:—When it is desired to *increase* the normal nutrition of a tissue or organ, we must not only supply it freely with the special materials requisite for its development, growth, and repair, but at the same time call upon it for the performance of its normal functions—over-fed idleness insures morbid nutrition, not healthy life.

In the next place I will give you a few *Rules* which may assist you in carrying out these general principles.

Rule 1.—When the power of appropriating any essential ingredient of a normal diet is lost to the organism, the lost function must be substituted by some artificial process, or the ingredient in question must be withdrawn from the diet till the normal function is restored. In obedience to this rule we administer pancreatic emulsions of fat to patients, who have lost the power of assimilating fat without this artificial assistance, while we adopt all practicable means of restoring the normal function.

Rule 2.—Is inseparable from the first, and it is this: —No essential of a normal diet must be withdrawn, without an attempt being made either to supply to the organism in some other way the ingredient of which it is deprived, or to suspend those functions which call for a supply of this ingredient. Thus, to take a simple illustration:—Suppose the power of digesting meat to be lost through a deficient secretion of gastric juice, meat must be withdrawn from the diet till the lost function is restored, or else an artificial digestive fluid must be introduced; or if it is impossible by these means to maintain the digestion of meat, the physiological ingredient of meat must be supplied in the form of some albuminoid solution; or, finally, if this cannot be done, then those functions which principally waste the albuminoid tissues of the body must be placed as far as possible in a state of rest, muscular action must be suspended until the function is restored.

Rule 3.—If an undue waste of any elements of normal nutrition is found to be going on in the organism, and the means remain of appropriating those elements from the food, they must be supplied in the food in quantities as much in excess of those proper to the normal diet of health as will be sufficient to supply the waste, until it is stopped, This also may be illustrated by a very simple example. In Bright's disease of the kidney there is no loss of the power to appropriate the albuminoids from the food, whereas a constant loss of albumen is going on through the kidneys, which must be met by proportionate increase of the albuminoids in the diet. But in following this rule, in this particular case, it will be necessary to observe the precautions which I mentioned when speaking of the fourth general principle.

Rule 4.—When through any defect in the organism, the elements of a normal diet are lost to nutrition if presented in the usual forms, those forms must be changed; but care must be taken that in the altered form all the essential elements of a normal diet are supplied in their proper quantities and proportions. Nothing can illustrate this better than the use of milk as a substitute for solid or mixed foods in diarrhœa or sickness.

Rule 5.—Has to deal with more complicated difficulties. If such a defect exists in the organism that *some* of the essentials of a normal diet was misappropriated, so that the organism is deprived of one or more of the normal elements of nutrition, and at the same time a disease is constituted out of the misappropriated food, then we have a double duty in interfering with the diet. First, the source of the disease must be stopped by withdrawing that part of the diet out of which it is constituted; and, secondly, the elements of nutrition thus removed must be supplied by some other means or in some other form.

Thus, in diabetes, the saccharine and amylaceous elements of the diet are misappropriated; they do not serve their normal function of supplying carbon for the evolution of heat, and by passing off through the kidneys they constitute an exhausting disease. It is necessary, therefore, to stop the source of this disease by cutting off the saccharine and amylaceous ingredients of the diet till normal nutrition is restored. But, in the meantime, as carbon must be obtained by some means, it is taken from the fat stored up in the body so long as that lasts, and when it is gone from the albuminoid tissues themselves, till the whole organism is disintegrated; unless at the same time that we cut off the starch and sugar, we increase the quantity of *fat* supplied in the food as much in excess of the proportion proper to a normal diet as shall fully supply the demand.

The modern dietetic treatment of diabetes may be taken as a good example of the way in which increased knowledge of the nature of disease and of the physiology of food enables us to act under what I have called the fifth principle of diet, viz., to avoid any unnecessary reduction in the number and variety of the forms in which food can be taken. In former days the poor parched diabetic was forbidden to drink water, lest he should increase his flow of urine; now we are able to let him quench his thirst as much as he pleases, so that he takes nothing which contains starch or sugar; and again, by preparing his articles of food in such a manner as to exclude the injurious ingredients, and by selecting those which are known to contain them in the smallest quantities, or not to contain them at all, we are able to present the diabetic with a fairly tempting and varied diet, so that he is able to keep to it for months and years with comparatively little difficulty.

DIETS FOR CONSUMPTION.

In these Diets for Consumption it is assumed that no fat is assimilated except that artificially pancreatized.

In Table I. The required amount of carbon is supplied by an excess of carbo-hydrates.

In Table II. The required amount of carbon is supplied by an excess of albuminoids.

In Table III. The amount of carbon is kept low, because it is only intended as a temporary diet, to be used during periods of rest in a warm room. The arrowroot and some of the fat of the milk are pancreatized by mixture with the "pancreatic emulsion."

DIETS FOR CONSUMPTION.—TABLE I.—CARBO-HYDRATE.

Food for 24 hours	Oz.	Carbon from	
		Nitrogenous.	Non-Nitrogenous.
Cooked Meat	6	0·732	0·420
Bread	10	0·540	1·930
Potatoes	3	0·072	0·760
Sugar	2	..	0·848
Milk 20 fluid ozs.	2½	0·540	0·900
Liebig's Foods for Infants	2	0·162	0·558
Farinaceous Foods	6	0·160	2·190
Fermented Liquors*	..	..	1·000
Pancreatic Emulsion	2	..	0·740
Totals	37½	2·206	9·340
Deduct Carbon from Non-pancreatized Fats as waste	..	..	0·945
Total available Carbon	..	..	10·607

* Either—Half a pint (Imperial) of Port, Sherry, or Marsala; or

DIETS FOR CONSUMPTION.—TABLE II.—ALBUMINOID.

With this Diet Hydrochloric Acid and Pepsine should be given to assist in digesting the very large quantity of Plastic Matter.

Food for 24 hours.	Oz.	Carbon from Non-Nitrogenous.	Nitrogenous.
Cooked Meat	8	0·976	0·560
Pigeon or Game	6	0·740	0·090
Dried Fish	3	0·710	0·035
Cheese	1	0·166	0·200
Vermicelli	3	0·777	0·516
Bread	4	0·220	0·770
Rice or Arrowroot	6	0·160	2·190
Sugar	3	. .	1·270
Milk, 20 fluid oz.	2½	0·540	0·900
Green Vegetables	6	0·030	0·204
Fermented Liquors*	. .	. .	1·000
Pancreatic Emulsion	2	. .	0·740
TOTALS	43½	4·319	8·475
Deduct Carbon from Non-pancreatized Fats as waste	. .	. .	1·410
Total available Carbon	. .	. .	11·384

One Pint of Burgundy, Claret or other similar Wine; or, One Pint of good Ale or Stout; or, a quarter of a pint of Rum, Whisky or Brandy, diluted with one pint of water.

---

*Either—Half a pint (Imperial) Port, Sherry, or Marsala; or, One Pint of Burgundy, Claret, or other similar Wine; or, One Pint of good Ale or Stout; or, a quarter of a Pint of Rum, Whisky or Brandy, diluted with one pint of water.

DIETS FOR CONSUMPTION.—TABLE III.—FLUID DIET.

	Oz.	Carbon from Nitrogenous.	Carbon from Non-Nitrogenous.
Milk 78 fluid oz.	10	2·106	3·510
Arrowroot	6	0·160	2·190
Pancreatic Emulsion	2	. .	0·740
TOTALS	18	2·266	6·440

This diet is to be given as follows:—

    8 ozs. of Milk and 1 oz. of Arrowroot every 4 hours (6 times in 24 hours) for 24 hours.

    10 ozs. of Milk and 1 oz. of Arrowroot every 4 hours for 24 hours.

    12 ozs. of Milk and 1 oz. of Arrowroot every 4 hours for 24 hours.

    13 ozs. of Milk and 1 oz. of Arrowroot every 4 hours for 24 hours.

    The last quantity is to be continued until solid diet can be borne by the stomach.

    One-third of an oz. of Pancreatic Emulsion is to be mixed with a little water, or with a portion of the milk, and given directly after each dose of Arrowroot and Milk, not mixed with the whole bulk.

DIET FOR DIABETES.

Table IV. In this Diet starch and sugar are reduced to a minimum and fat and albuminoids are given in their place, some of the fat being introduced in the form of Pancreatic Emulsion to assist in its assimilation.

TABLE IV.—DIET FOR DIABETES.

With this diet Hydrochloric Acid and Pepsine should be given to assist in digesting the very large quantity of Plastic Matter.

Food for 24 hours.	Dry oz.	Saccharine-	Total Carbon.	Carbon from Nitrogenous	Carbon from Non-Nitrogenous	Carbon from Saccharine portions of Non-nitrog's.
Cooked Meat and Poultry	8	..	1·536	0·976	0·560	..
Cooked Pigeon or Game.	6	..	0·830	0·740	0·090	..
Dried Fish (Haddock) .	3	..	0·745	0·710	0·035	..
Cheese . . .	1	0·024	0·366	0·166	0·200	..
Van Abbott's Gluten Bread or Gluten Vermicelli .	6	1·148	2·438	1·928	0·510	0·510 ⎫
Green Vegetables* .	3	0·234	0·117	0·015	0·120	0·102 ⎬ 0·612
Fermented Liquors† (Brandy or Whisky) 5 fluid ozs.	..	..	1·000	..	1·000	..
Butter (pure) . .	2	..	1·480	..	1·480	..
Pancreatic Emulsion .	1½	..	0·555	..	0·750	..
Eggs (2) . . .	3	..	0·260	..	0·260	..
Bacon . . .	3	..	1·620	0·135	1·485	..
Totals . . .	36	1·406	10·947	4·670	6·472	..
Carbon from Saccharine to be deducted as waste	..	..	0·612	..	..	..
Total Carbon available	..	..	10·335	..	..	..

Salt to taste. Water as much as required by thirst. Tea without sugar, with a slice of lemon peel in it.

*Green vegetables permitted—Cress, Celery, Endive, Greens, Lettuce, Mustard, Spinach, Water-cress.

†The following Wines may take the place of Spirits (for equivalent quantities see Alcohol Table), Claret, Moselle (still) Rhine Wine, Manzanilla, Greek (St. Elie) very dry Amontillado Sherry.

NUTRITIVE ENEMATA.

IN the experiments upon digestion, performed by Czerny and Latschenberger (Virchow's *Archiv*, Band lix., Heft ii.), comparative trials were made between the materials introduced into the large intestine, and the same acted upon outside of the body with the secretion obtained from the mucous membrane. Portions of hard-boiled white of egg and shreds of fibrine remained unchanged, preserving the sharpness of their angles and borders, when exposed to the action of the mucus at a temperature of 100° for two or three hours. No emulsion could be obtained by shaking up olive oil and the mucus; and no conversion of starch into sugar could be thus produced, even after the lapse of several hours. Similar cubes of hard-boiled white of egg were retained in the rectum in small perforated capsules for no less a period than ten weeks, and yet on withdrawal exhibited no indication of any digestive action. Experiments made with soluble albumen in like manner showed that the large intestine of man exerts no digestive action upon it.

Other experiments, made with a view of determining the absorptive capacity of the portion of intestine under observation, and which, as before stated, was estimated at about 240 square centimetres, showed that in the course of seven hours the quantity of water that could be taken up was from 617 to 772 grains. They showed also that although the intestinal juices exerted no digestive action on albumen, and no emulsifying action on fat, yet that the walls of the intestine were capable of *absorbing* both albumen when introduced in the soluble form, and oil if it had been previously emulsified. The quantity of soluble albumen absorbed was always proportionate to the time. Any irritation applied to the intestine checked the process of absorption, and, if violent, stopped it altogether. Raw white of egg was found to be an unfavorable form for absorption. The best mode of preserving life by means of injection is often an important subject of consideration, especially in cases of cancer of the intestines; and these experiments accord with the observations and recommendations of LEUBE, that whilst comparatively little benefit can be obtained from the injection of the raw material of our ordinary diet, considerable quantities can be absorbed, and much improvement can be produced in the strength and health of the patient, if the substances have been previously subjected to operations by which they are partially digested—as, for instance, if fat be emulsified, if albumen be reduced to the soluble state, and if starch have been converted into glycose.

DIRECTIONS FOR SPECIAL ARTICLES OF DIET IN DISEASE.

BEEF TEA should not be boiled, and should not be strained through a fine sieve or muslin. It should be made as follows. Take of Rumpsteak, free from fat and minced, 1 lb., *cold* water 1 pint, a pinch of salt. Put them into a jar and tie it down. Place the jar in a saucepan of *cold* water, raise this water slowly to boiling for two hours. Remove the jar and strain its contents through a *very coarse* sieve so that all finely powdered sediment may run through. Then pass a piece of bread over the surface to remove any fat that may float upon it.

LIEBIG'S EXTRACT OF MEAT and other similar preparations. It is important to bear in mind that *these contain very little, if any, nourishment properly so-called;* that is to say, they contain no plastic material, no fat, no saccharine matter. Their principal virtues belong to the class of stimulants and blood-tonics When mixed with water, they are excellent menstrua in which to administer nutritive materials, such as eggs, bread, oatmeal, corn-flour, vermicelli; but without such additions they are quite incapable of supporting life for any length of time. Baron Liebig's own writings support this statement. Unless these facts are borne in mind a patient may easily be starved unintentionally.

WHITE OF EGG differs from the yolk principally in containing no fat. On this account it is often better borne by bilious persons. Yolk of Egg contains 29.8 per cent. of fat; when the stomach can bear it, therefore, it is a more complete nutriment than the white. But White of Egg beaten up *in milk* answers every purpose.

EGGS for the sick should be either raw, or very lightly boiled.

SPECIAL RESTORATIVE. I have found the following to be a most efficient restorative food, and the mixture is agreeable to most palates.

>   New milk (cold) 4 parts.
>   Beef tea (cold) 3 parts.
>   Brandy (pale) 1 part.

If no other food is taken, about 5 fluid-ounces (a quarter of a pint Imperial) should be given every 2 hours, or half that quantity every hour. It should be sucked out of a *syphon* infant's-feeding bottle, not drunk out of a spoon or cup. When desirable this food may be gradually thickened by the addition of boiled corn-flour or other farinaceous articles, and one egg may be well beaten up in each half pint. The flavor may be varied by adding different spices.

SPECIAL NUTRITIVE. Beat up an egg, both white and yolk, quite smooth and free from stringy particles, stir it well into half a pint of hot milk in which enough arrowroot has been boiled to make it about as thick as cream; add a wineglassful of sherry or tablespoonful of pale brandy, five grains of Pancreatine powder (Savory and Moore's), and some fresh nutmeg; mix all thoroughly by pouring from cup to cup. On this food alone, repeated every four hours, a patient can be well supported for a considerable time.

COCOA AND EGG. Beat up an egg, both white and yolk, quite smooth and free from stringy particles, stir it into half a pint of hot milk, and then add a teaspoonful of soluble cocoa, previously liquefied with a little of the milk. This forms an excellent breakfast, easily taken by those who cannot eat in the early part of the day.

INVALID SOUP. The following Invalid Soup has proved extremely useful in a large number of cases, and since I first published the recipe in 1864 it has been usually kept nicely prepared by Mr. Donges, Confectioner, Gower Street, W. C.

Gravy beef 1 lb., scragg of mutton 1 lb., isinglass 2 oz., vermicelli 3 oz., mushroom ketchup 3 tablespoonfuls, corns of allspice 24, sage a sprig, cold water 3 quarts; put the isinglass and the meat cut small into the cold water, gradually boil, skim well, and then add the other ingredients; simmer four or five hours till reduced to one quart; strain through a fine hair sieve, and carefully remove all fat; add salt to the taste. This may be taken cold as a jelly, or warm as a soup. Calf's-foot may be used instead of isinglass when procurable; and when allowable a little *solution* of cayenne pepper should be added; and the taste may be varied by the addition of a little Worcester, or other wholesome sauce.

COMBINATIONS OF ALIMENTARY PRINCIPLES IN NEARLY EXACT NORMAL PROPORTIONS.

a. Flour 4 oz., sugar 1¼ oz., suet ¾ oz., milk ¾ pint Imperial, 1 egg.—This will make a good pudding, or it may be given in any other form desired; with the addition of a little cress and salt and water it forms a complete diet, upon a sufficient quantity of which a person can live healthfully for an indefinite length of time without any other food.

b. The same may be said of the following.—Rice 3 oz., sugar 1 oz., 2 eggs, butter ½ oz., milk ¾ pint (Imperial), water as much as is sufficient to boil the rice in.

c. Suet ¼ lb., flour 1 lb., water 13 oz. These quantities when boiled yield 2 lbs. of pudding.

PORT WINE JELLY. Take of port wine 1 pint, isinglass 1 oz., sugar 1 oz., put the isinglass and sugar into ¼ pint of water, warm till all is dissolved, then add the wine, strain through muslin and set to jelly. (An excellent way of giving port wine.)

Another form, firm enough to carry in the pocket cut up in cubes, may be made as follows:

Take isinglass and gum Arabic of each an ounce, dissolve in a pint of port wine over a slow fire; sweeten with fine sugar to the taste, and after straining through a fine sieve, grate in a small nutmeg. Take about a cubic inch when feeling weak or low.

SUET AND MILK. Put a tablespoonful of shredded beef suet into ½ pint of fresh milk, warm it sufficiently to completely melt the suet, then skim it, pour it into a *warm* glass or cup, and drink it before it cools. If there is any difficulty in digesting the suet add 5 gr. of Pancreatine powder. (Savory and Moore's.)

MILK WITH RUM, BRANDY, OR WHISKEY. Put one tablespoonful of Rum, Brandy, or Whiskey into half a pint of new milk, and mix well by pouring several times from one vessel to another. "Bilious" persons should heat the rum before adding it to the milk.

NUTRITIVE ENEMATA. When nutriment is given in enemata the quantity should not exceed from 2 to 4 oz., and the temperature should be about 80°.

The bowel should be first washed out with half a pint of warm water. An elastic bottle holding the required quantity is better for nutritive enemata than the ordinary enema syringe. They should be given while the patient is lying on the back with the hips raised on a pillow.

The following constitutes a most important means of preserving life when food cannot be given by the stomach.

> Take of cooked beef or mutton finely grated ¼ lb.
> Pancreatic Emulsion (Savory and Moore's) 1 oz.
> Pancreatine powder (Savory and Moore's) 20 grains.
> Pepsine (Porci) 20 grains.

Mix the whole in a warm mortar quickly, and add Brandy one table-spoonful and enough warm water to bring the mixture to the consistence of Treacle. Inject from an elastic Enema bottle, as quickly after the mixture is made as possible, and let it be retained.

NUTRITIVE MIXTURE. When a patient will take medicine but no food.

> Liebig's Extract of Meat a tea-spoonful.
> Lœflund's Liebig's Extract of Malt a tea-spoonful.
> Tincture of Capsicum one drop.
> Compound spirit of Horseradish a tea-spoonful.
> Water 2 table-spoonfuls. Mix well in a mortar.

To be given every 3 or 4 hours. This will often bring back the desire for food.

INDEX.

	PAGE.
Abdominal breathing	17
Aconite, use of	152
Alcoholic beverages in consumption	206
Ammonia, Ses carbonate of	135, 151, 169
Ammoniacum in cough	155
Ammonium, chloride of	156, 169
" bromide of	179
Amphoric resonance	29
Amyl-nitrite in asthma	157
Antimony in cough	152
Artificial respiration in cough	162
Asthma, its relation to Emphysema, 63; treatment 156; influence on cough	178
Asthmatic subjects, bronchitis of,	153, note.
Atomized fluids, use of	165
Auditory canal, sympathy between and larynx	101
Auscultation, terms used in, 19; rules regarding	21
Avoidance of colds	142
Beef tea, how made	211
Benzoin in cough	156, 168
Bimeconate of morphia in cough	181
Bismuth in cough	180
Blisters in cough	170
Blood, spitting of—*see* hæmoptysis.	

(217)

	PAGE.
Borax in catarrh	183
Bromide of ammonium	179
Bronchi, spasmodic contraction of	156
Bronchial catarrh	117
Bronchitis, in winter cough 49, 55; collapsed lung from, 123; of asthmatic subjects, 153; turpentine in	169
Bruit de pôt fêlê	30
Camphor in catarrh	183
Cannabis indica	179
Carbon in food	205
Catarrh, post-nasal 89, 182; naso-pulmonary, 114; early treatment of, 132; epidemic	136
Cavernous sounds, value of	42
Chest, movements of in disease	17
Chloral in cough	180
Chloride of ammonium in cough	156
Climatic treatment of colds	172
Cocoa and egg	213
Cod liver oil, digestion of	192
Cold, how to stop a, 134; how to avoid	143
Collapse of the lung	57, 123
Congestion, bronchial, mode of relieving	170
Conium in coughs	180
Consumption, early signs of	37
" management of	185
" pancreatic emulsion in	190
" true first stage of	37, 191
" diets for	205
Copaiba in cough	155
Coryza, how distinguished from post-nasal catarrh	94
Cough, ear, 97; natural course of neglected, 112; therapeutic resources of,	150

Index.

	PAGE
Counter-irritation in coughs	169
Cubebs in cough	155
Datura tatula in asthma	156
Diabetes, diet of	203, 207, 208

Diagnosis of early phthisis, 37; of narrowed air-passages . . 78
Diathesis, the treatment of 140, 176
Diet in disease, 193; for consumption, 205; for diabetics, 207;
    special recipes for, ... 24
Digitalis, use in cough 181
Disease, diet in, 193; physiology of 194
Dropsy, from winter cough 123
Ear cough 97
Early signs of consumption 37
Early treatment of catarrh 132
Eggs, in diet 212
Emphysema, in winter cough, 49, 55, 126; how produced, 56;
    phenomena of, explained, 75; forms of . . . 174
Eucalyptus globulus in cough . . . . . 160, note.
Enemata, nutritive 209
Expiratory movement, of undue length 17
Extract of meat, Liebig's 211
Fat in food 203
Fleming, Dr., experiments with aconite . . . 153
Fox, Dr. Cornelius B., on ear cough 98
Fremitus 19
Friction sound 31
Friederichshall water 159
Gout, as a complication in cough 158, 159
Hæmoptysis, import of as a symptom, 45; arrest of . 166, note.
Heart, dilatation of 122
Hydrocyanic acid in cough 180

	PAGE.
Indian hemp, *see* Cannabis.	
Influenza, how to check	136
Inhalations in coughs	165
Interlobular emphysema	59, 64, 75
Invalid soup	213
Ipecacuanha, in cough	153
Liebig's extract of meat	211
Lobelia inflata, in asthma, 157; in cough	179
Lobular emphysema	74
Localized rest in lung disease	186
Lozenge, medicated	183
Lung, collapse of	57, 123
Lungs, position of in the chest, 24; elasticity of	128
Lungs, rest to the	185
Lung-splint	185
Meat, Liebig's extract of	211
Mensuration of chest in disease	18
Mercury, bichloride, in coughs	159
Metallic tinkling sound	28
Metastasis of skin diseases	160
Milk and suet	215
Milk punch	215
Morphia, bimeconate of	181
Naso-pulmonary catarrh, 114; climatic treatment of	173
Neglected cough, course of	112
Nitrite of amyl	157
Nutritive enemata	209, 215
Nutritive, special	213
Nutritive mixture	216
Olibanum in cough	155
Out-door life, value of	149

	PAGE.
Palpation of the chest	19
Pancreatic emulsion in consumption	190
Pancreatine	190
Passages, diagnosis of narrowed air	78
Phthisis, diagnosis of early	37
Pitch, modifications of, 78; as diagnostic sign	88
Pitch plasters	171
Port wine jelly	214
Post-nasal catarrh, diagnosis, 89; treatment of	182
Poultices, use of	172
Questions for diagnosis	52
Respiration, artificial	162
Respirators, value of	147
Rest to lungs, 185; in disease	198
Restorative, special	212
Rhonchus	26
Saline aperients, use of	159
Senega in cough	154
Serpentaria in Cough	155
Skin diseases, metastasis of	160
Sleeping rooms, temperature of	148
Snuff, medicated	183
Sounds, cavernous, value of	42
Soup, invalid	213
Spasmodic asthma and emphysema, 62; inspiratory sound in, 85; treatment of,	156, 179
Special nutritive	213
Special restorative	212
Spray producers	166
Squills in cough	153
Stimulants in food	206
Stramonium in cough	158
Strychnia in emphysematous cough	160

	PAGE
Succussion sound, how imitated	28
Suet and milk	215
Therapeutics of cough	150
Timbre as an attribute of sound	79
Tonics in cough	160
Transmitted sounds	87
Tubercle, first signs of deposit of, 40; what it is, 48; often associated with emphysema,	89
Turpentine in chronic bronchitis	169
Tympanitic resonance	29
Vaporizers	167
Vertigo from aural irritation	103
Vesication in coughs	170
Vomiting from aural irritation	111
Wet feet, dangers of	145, 146
Whooping cough, collapsed lung in, 123; cough of,	127
Winter coughs, remarks on, 49; course of, 112; pathological conditions in,	121

www.ingramcontent.com/pod-product-compliance
Lightning Source LLC
Chambersburg PA
CBHW031815230426
43669CB00009B/1151